The **BEST** **WRITING** on **MATHEMATICS**

2019

The **BEST** **WRITING** on **MATHEMATICS**

2019

Mircea Pitici, Editor

PRINCETON UNIVERSITY PRESS
PRINCETON AND OXFORD

To Vickie Kearn,
for our ten years of collaboration on this series
and for her decades-long enthusiasm
in bringing to readers excellent books on mathematics

Contents

Color illustrations follow page 80

Introduction

MIRCEA PITICI

This volume is the tenth in our annual series of *The Best Writing on Mathematics*. For me, the editor, the series is a continuous enterprise of reading and curating the vast literature on mathematics (published in research journals, books, magazines, and online) fragmented yearly by our time-measuring conventions. I encourage the readers who want to enjoy fully the results of this endeavor and the reviewers who judge its merits to consult the series in its entirety, not only the current (or any other one) volume. In the introductions to the previous volumes in the series I detailed the procedure we follow to reach the selection; I will not repeat it here. The ten volumes available so far contain more than 200 pieces.

I selected the contents of this anthology from mostly 2018 materials. That, and the circumstance of appearing here together, might be the only characteristics common to the pieces in the book. The collection is once again purposefully eclectic, guiding the reader toward a non-dogmatic understanding of mathematics—panoramic, diverse, open to interpretive possibilities, and conducive to pertinent connections. Mathematics is unique in the asymmetry between its apparent singularity of method, circumscribed to the rigors of syllogistic reasoning, and the disconcerting multiplicity of its reverberations into other domains. Somehow disturbing is yet another asymmetry, between mathematicians' attempts to establish unambiguous, clear statements of facts, and the wide range of potentially harmful applications to which mathematical notions, methods, results, and conclusions are used, indiscriminately and independent of context. Discerning the proper use of mathematics in applications from the improper one is no trifling matter, as some of the contributors to this series of anthologies have pointed out over the years.

Before I present the articles you can find in the book, I remind you that this is not only an anthology of intriguing and stimulating readings but also a reference work meant to facilitate an easy orientation into the valuable literature on mathematics currently published in a broad array of sources. The list of books I give in this introduction and the additional lists grouped at the end of the book under the title "Notable Writings" are the main components of the bibliographic aspects of the series.

Overview of the Volume

To start the selection, Moon Duchin explains that the Markov chain Monte Carlo method, a geometric-statistical approach to the analysis of political districting, guards against the worst of many possible abuses currently taking place within elective political processes.

Theodore Hill describes the recent history of the fair division of a domain problem, places it in wider practical and *im*practical contexts, and traces the contributions of a few key mathematicians who studied it.

Paul Campbell examines some of the claims commonly made on behalf of learning mathematics and finds that many of them are wanting in the current constellation of teaching practices, curricula, and competing disciplines.

Roice Nelson introduces several puzzles whose ancestry goes back to the famous cube invented and commercialized by Ernő Rubik.

Kokichi Sugihara analyzes the geometry, the topology, and the construction of versatile three-dimensional objects that produce visual illusions when looked at from different viewpoints.

Kevin Hartnett traces the recent developments and the prospects of mathematical results that establish mirror symmetry between algebraic and simplectic geometry—an unexpected and only partly understood correspondence revealed by physicists.

James Propp presents a fresh approach to problems of discrete probability and illustrates it with examples of various difficulties.

Neil Sloane details some of the remarkable numerical sequences he included in the vast collection of integers he has organized and made available over the past several decades.

Alessandro Di Bucchianico, Laura Iapichino, Nelly Litvak, Frank van der Meulen, and Ron Wehrens point out specific theoretical advances in various branches of mathematics, which have contributed powerful applications to recent technologies and services.

Toby Cubitt, David Pérez-García, and Michael Wolf tell us how they explored the connections between certain open questions in quantum physics and classical results on undecidable statements in mathematics formulated by Kurt Gödel and Alan Turing.

Jeremy Avigad places in historical context and illustrates with recent examples the growing use of computation, not only in proving mathematical results but also in making hypotheses, verifying them, and searching for mathematical objects that satisfy them.

With compelling examples and well-chosen arguments, Reuben Hersh makes the case that mathematics is pluralistic on multiple levels: in content, in philosophical interpretation, and in practice.

Mary Leng subtly defends a position highly unpopular among mathematicians and in a small minority among the philosophers of mathematics, namely, the thesis that certain mathematical statements are questionable on the ground that they imply the existence of objects that might not exist at all—for instance abstract numbers.

Tiziana Bascelli and her collaborators (listed in alphabetical order), Piotr Błaszczyk, Vladimir Kanovei, Karin U. Katz, Mikhail G. Katz, Semen S. Kutateladze, Tahl Nowik, David M. Schaps, and David Sherry, discuss an episode of 17th-century nonstandard analysis to argue that clarifying both the historical ontology of mathematical notions and the prevalent procedures of past times is essential to the history of mathematics.

Noson Yanofsky invokes two paradoxes from the realm of numbers and a famous result from the mathematical theory of complexity to speculate about their potential to inform our understanding of daily life.

Andrew Gelman recommends several practices that will make the communication of statistical research, of the data, and of their consequences more honest (and therefore more informative) to colleagues and to the public.

Michael Barany narrates a brief history of the early Fields Medal and reflects on the changes that have taken place over the decades in the award's stated aims, as well as in the manner in which awardees are selected.

To conclude the selection for this volume, Melvyn Nathanson recalls some originalities of one of the most peculiar mathematicians, Paul Erdős.

More Writings on Mathematics

Besides the pieces included in the anthology, every year I suggest other readings, offering a quick overview of books that came to my attention recently, loosely grouped in several thematic categories. This list is lacunary; it consists only of books I consulted myself, either in the two excellent libraries accessible to me (at Syracuse University and Cornell University—thank you!) or sent to me by authors and publishers. Full references are included at the end of the introduction.

A direct marketplace competitor to this volume deserves a special mention: the outstanding anthology *The Prime Number Conspiracy*—in which the editor Thomas Lin included pieces previously published by the online magazine *Quanta*. An original celebration of fertile problems in mathematics, abundantly supplemented with theoretical introductions (sometimes quite technical for the general reader) is *100 Years of Math Milestones* by Stephan Ramon Garcia and Steven Miller.

Among books exploring the presence of mathematics in daily life, human activities, and the natural world, some titles are *The Beauty of Numbers in Nature* by Ian Stewart, *Humble Pi* by Matt Parker, *Weird Math* by David Darling and Agnijo Banerjee, *Outnumbered* by David Sumpter, and *The Logic of Miracles* by László Mérő. A somehow more technical expository book but widely accessible is *Exercises in (Mathematical) Style* by John McCleary.

Some books about data, statistics, and probability are *Using and Interpreting Statistics in the Social, Behavioral, and Health Sciences* by William Wagner III and Brian Joseph Gillespie, *Narrative by Numbers* by Sam Knowles, *The Essentials of Data Science* by Graham Williams, and *The Politics of Big Data* edited by Ann Rudinow Sætnan, Ingrid Schneider, and Nicola Green.

Two interesting books about mathematics in past cultures are *Scale and the Incas* by Andrew James Hamilton and *Early Rock Art of the American West* by Ekkehart Malotki and Ellen Dissanayake. Other books on the history of mathematics or on the role of mathematics in past societies are *A History of Abstract Algebra* by Jeremy Gray, *Reading Popular Newtonianism* by Laura Miller, *A People's History of Computing in the United States* by Joy Lisi Rankin, *Exact Thinking in Demented Times* by Karl Sigmund, and *Calculated Values* by William Deringer. Two recent histories of calendars, including the mathematics of calendars, are *Scandalous*

Error by Philipp Nothaft and *Calendrical Calculations* by Edward Reingold and Nachum Dershowitz. Two biographies of mathematicians are *The Young Descartes* by Harold Cook and the autobiographical *The Shape of a Life* by Shing-Tung Yau and Steve Nadis.

Connections with other disciplines and applied mathematics can be found, for instance, in the well-illustrated *Math Art* by Stephen Ornes, *Figuring Fibers* edited by Carolyn Yackel and sarah-marie belcastro, *What Shape Is Space?* by Giles Sparrow, *Strategies for Quantitative Research* [in archaeology] by Grant McCall, *The Oxford Handbook of Computational Economics and Finance* edited by Shu-Heng Chen, Mak Kaboudan, and Ye-Rong Du, and *Essential Discrete Mathematics for Computer Science* by Harry Lewis and Rachel Zax. Specifically on mathematics, physics, and astronomy and other sciences are *Lost in Math* by Sabine Hossenfelder, *Welcome to the Universe* by Neil deGrasse Tyson, Richard Gott, and Michael Strauss, *Anxiety and the Equation* by Eric Johnson, and *Alice and Bob Meet the Wall of Fire* edited by Thomas Lin.

Of the many books on mathematics education, recently I have seen *Brainball* by Mickey Kolis and Cassandra Meinholz and *Modeling Mathematical Ideas* by Jennifer Suh and Padmanabhan Seshaiyer. Of wider scope are *Education by the Numbers and the Making of Society* edited by Sverker Lindblad, Daniel Pettersson, and Thomas S. Popkewitz and *The Nature and Development of Mathematics* edited by John Adams, Patrick Barmby, and Alex Mesoudi.

In the philosophy of mathematics and related essays, some recent books are *Abstraction and Infinity* by Paolo Mancosu, *What Is a Mathematical Concept?* edited by Elizabeth de Freitas, Nathalie Sinclair, and Alf Coles, *Naturalizing Logico-Mathematical Knowledge* edited by Sorin Bangu, *Making and Breaking Mathematical Sense* by Roi Wagner, *Mathematics and Its Applications* by Jairo José da Silva, *Explanation beyond Causation* edited by Alexander Reutlinger and Juha Saatsi, *Probabilistic Knowledge* by Sarah Moss, *Reality and Its Structure* edited by Ricki Bliss and Graham Priest, *Truth, Existence, and Explanation* edited by Mario Piazza and Gabriele Pulcini, and *Robustness Tests for Quantitative Research* by Eric Neumayer and Thomas Plümper. In logic, *Formal Theories of Truth* by J. C. Beall, Michael Glanzberg, and David Ripley, *The Significance of the New Logic* by Willard Van Orman Quine, *Logic, Language, and the Liar Paradox* by Martin Pleitz, *Logic from Kant to Russell* edited by Sandra Lapointe, and *Cycles and Social Choice* by Thomas Schwartz.

Broadly interdisciplinary are *The Mathematical Imagination* by Matthew Handelman, *Faith across the Multiverse* by Andy Walsh, *The Great Rift* by Michael Hobart, and *The Experimental Side of Modeling* edited by Isabelle Peschard and Bas van Fraassen.

ᏇᏇᏇ

I hope that you, the reader, will enjoy reading this anthology at least as much as I did while working on it. I encourage you to send comments, suggestions, and materials I might consider for (or mention in) future volumes to Mircea Pitici, P.O. Box 4671, Ithaca, NY 14852; or electronic correspondence to mip7@cornell.edu.

Books Mentioned

Adams, John W., Patrick Barmby, and Alex Mesoudi. (Eds.) *The Nature and Development of Mathematics: Cross Disciplinary Perspectives on Cognition, Learning and Culture.* Abingdon, U.K.: Routledge, 2018.

Bangu, Sorin. (Ed.) *Naturalizing Logico-Mathematical Knowledge: Approaches from Philosophy, Psychology and Cognitive Science.* New York: Routledge, 2018.

Beall, J. C., Michael Glanzberg, and David Ripley. *Formal Theories of Truth.* Oxford, U.K.: Oxford University Press, 2018.

Bliss, Ricki, and Graham Priest. (Eds.) *Reality and Its Structure: Essays in Fundamentality.* Oxford, U.K.: Oxford University Press, 2018.

Chen, Shu-Heng, Mak Kaboudan, and Ye-Rong Du. (Eds.) *The Oxford Handbook of Computational Economics and Finance.* Oxford, U.K.: Oxford University Press, 2018.

Cook, Harold J. *The Young Descartes: Nobility, Rumor, and War.* Chicago: Chicago University Press, 2018.

da Silva, Jairo José. *Mathematics and Its Applications: A Transcendental-Idealist Perspective.* Cham, Switzerland: Springer International Publishing, 2017.

Darling, David, and Agnijo Banerjee. *Weird Math: A Teenage Genius and His Teacher Reveal the Strange Connections between Math and Everyday Life.* New York: Basic Books, 2018.

de Freitas, Elizabeth, Nathalie Sinclair, and Alf Coles. (Eds.) *What Is a Mathematical Concept?* New York: Cambridge University Press, 2017.

Deringer, William. *Calculated Values: Finance, Politics, and the Quantitative Age.* Cambridge, MA: Harvard University Press, 2018.

Garcia, Stephan Ramon, and Stephen J. Miller. *100 Years of Math Milestones: The Pi Mu Epsilon Centennial Collection.* Providence, RI: American Mathematical Society, 2019.

Gray, Jeremy. *A History of Abstract Algebra: From Algebraic Equations to Modern Algebra.* Cham, Switzerland: Springer International Publishing, 2018.

Hamilton, Andrew James. *Scale and the Incas.* Princeton, NJ: Princeton University Press, 2018.

Handelman, Matthew. *The Mathematical Imagination: On the Origins and Promise of Critical Theory.* New York: Fordham University Press, 2019.

Hobart, Michael E. *The Great Rift: Literacy, Numeracy, and the Religion-Science Divide.* Cambridge, MA: Harvard University Press, 2018.

Hossenfelder, Sabine. *Lost in Math: How Beauty Leads Physics Astray.* New York: Basic Books, 2018.

Johnson, Eric. *Anxiety and the Equation: Understanding Boltzmann's Entropy.* Cambridge, MA: MIT Press, 2018.

Knowles, Sam. *Narrative by Numbers: How to Tell Powerful and Purposeful Stories with Data.* Abingdon, U.K.: Routledge, 2018.

Kolis, Mickey, and Cassandra Meinholz. *Brainball: Teaching Inquiry Math as a Team Sport.* Lanham, MD: Rowman & Littlefield, 2018.

Lapointe, Sandra. (Ed.) *Logic from Kant to Russell: Laying the Foundations for Analytic Philosophy.* New York: Routledge, 2019.

Lewis, Harry, and Rachel Zax. *Essential Discrete Mathematics for Computer Science.* Princeton, NJ: Princeton University Press, 2019.

Lin, Thomas. (Ed.) *Alice and Bob Meet the Wall of Fire: The Biggest Ideas in Science from* Quanta. Cambridge, MA: MIT Press, 2018.

Lin, Thomas. (Ed.) *The Prime Number Conspiracy: The Biggest Ideas in Math from* Quanta. Cambridge, MA: MIT Press, 2018.

Lindblad, Sverker, Daniel Pettersson, and Thomas S. Popkewitz. (Eds.) *Education by the Numbers and the Making of Society: The Expertise of International Assessments.* New York: Routledge, 2018.

Malotki, Ekkehart, and Ellen Dissanayake. *Early Rock Art of the American West: The Geometric Enigma.* Seattle: University of Washington Press, 2018.

Mancosu, Paolo. *Abstraction and Infinity.* Oxford, U.K.: Oxford University Press, 2016.

McCall, Grant S. *Strategies for Quantitative Research: Archaeology by Numbers.* Abingdon, U.K.: Routledge, 2018.

McCleary, John. *Exercises in (Mathematical) Style: Stories of Binomial Coefficients.* Washington, DC: Mathematical Association of America, 2018.

Mérő, László. *The Logic of Miracles: Making Sense of Rare, Really Rare, and Impossibly Rare Events.* New Haven, CT: Yale University Press, 2018.

Miller, Laura. *Reading Popular Newtonianism: Print, the* Principia, *and the Dissemination of Newtonian Science.* Charlottesville, VA: University of Virginia Press, 2018.

Moss, Sarah. *Probabilistic Knowledge.* Oxford, U.K.: Oxford University Press, 2018.

Neumayer, Eric, and Thomas Plümper. *Robustness Tests for Quantitative Research.* Cambridge, U.K.: Cambridge University Press, 2017.

Nothaft, C. Philipp E. *Scandalous Error: Calendar Reform and Calendrical Astronomy in Medieval Europe.* Oxford, U.K.: Oxford University Press, 2018.

Ornes, Stephen. *Math Art: Truth, Beauty, and Equations.* New York: Sterling Publishing, 2019.

Parker, Matt. *Humble Pi: A Comedy of Maths Errors.* London: Penguin Random House, 2019.

Peschard, Isabelle F., and Bas C. van Fraassen. (Eds.) *The Experimental Side of Modeling.* Minneapolis: University of Minnesota Press, 2018.

Piazza, Mario, and Gabriele Pulcini. (Eds.) *Truth, Existence, and Explanation.* Cham, Switzerland: Springer International Publishing, 2018.

Pleitz, Martin. *Logic, Language, and the Liar Paradox.* Münster, Germany: Mentis Verlag, 2018.

Quine, Willard Van Orman. *The Significance of the New Logic.* Cambridge, U.K.: Cambridge University Press, 2018.

Rankin, Joy Lisi. *A People's History of Computing in the United States.* Cambridge, MA: Harvard University Press, 2018.

Reingold, Edward M., and Nachum Dershowitz. *Calendrical Calculations: The Ultimate Edition.* Cambridge, U.K.: Cambridge University Press, 2018.

Reutlinger, Alexander, and Juha Saatsi. (Eds.) *Explanation beyond Causation: Philosophical Perspectives on Non-Causal Explanations.* Oxford, U.K.: Oxford University Press, 2018.

Sætnan, Ann Rudinow, Ingrid Schneider, and Nicola Green. (Eds.) *The Politics of Big Data: Big Data, Big Brother?* Abingdon, U.K.: Routledge, 2018.

Schwartz, Thomas. *Cycles and Social Choice: The True and Unabridged Story of a Most Protean Paradox.* Cambridge, U.K.: Cambridge University Press, 2018.

Sigmund, Karl. *Exact Thinking in Demented Times: The Vienna Circle and the Epic Quest for the Foundations of Science.* New York: Basic Books, 2017.

Sparrow, Giles. *What Shape Is Space? A Primer for the 21st Century.* London: Thames & Hudson, 2018.

Stewart, Ian. *The Beauty of Numbers in Nature: Mathematical Patterns and Principles from the Natural World.* Brighton, U.K.: Ivy Press, 2017 (first published in 2001).

Suh, Jennifer M., and Padmanabhan Seshaiyer. *Modeling Mathematical Ideas: Developing Strategic Competence in Elementary and Middle School.* Lanham, MD: Rowman & Littlefield, 2017.

Sumpter, David. *Outnumbered: From Facebook and Google to Fake News and Filter-Bubbles—The Algorithms That Control Our Lives.* London: Bloomsbury Sigma, 2018.

Tyson, Neil deGrasse, J. Richard Gott, and Michael A. Strauss. *Welcome to the Universe: The Problem Book.* Princeton, NJ: Princeton University Press, 2017.

Wagner, Roi. *Making and Breaking Mathematical Sense: Histories and Philosophies of Mathematical Practice.* Princeton, NJ: Princeton University Press, 2017.

Wagner, William E., III, and Brian Joseph Gillespie. *Using and Interpreting Statistics in the Social, Behavioral, and Health Sciences.* Los Angeles, CA: Sage, 2019.

Walsh, Andy. *Faith across the Multiverse: Parables from Modern Science.* Peabody, MA: Hendrickson Publishers, 2018.

Williams, Graham J. *The Essentials of Data Science: Knowledge Discovery Using R.* Boca Raton, FL: CRC Press, 2017.

Yackel, Carolyn, and sarah-marie belcastro. (Eds.) *Figuring Fibers.* Providence, RI: American Mathematical Society, 2018.

Yau, Shing-Tung, and Steve Nadis. *The Shape of a Life: One Mathematician's Search for the Universe's Hidden Geometry.* New Haven, CT: Yale University Press, 2019.

Geometry v. Gerrymandering

Moon Duchin

Gerrymandering is clawing across courtrooms and headlines nation-wide. The U.S. Supreme Court recently heard cases on the constitutionality of voting districts that allegedly entrenched a strong advantage for Republicans in Wisconsin and Democrats in Maryland but dodged direct rulings in both. Another partisan gerrymandering case from North Carolina is winding its way up with a boost from an emphatic lower court opinion in August. But so far, it has been impossible to satisfy the justices with a legal framework for partisan gerrymandering. Part of the problem, as former Justice Anthony Kennedy noted in a 2004 case, is that courts high and low have yet to settle on a "workable standard" for identifying a partisan gerrymander in the first place. That is where a growing number of mathematicians around the country think we can help.

Two years ago, with a few friends, I founded a working group to study the applications of geometry and computing to redistricting in the United States. Since then, the Metric Geometry and Gerrymandering Group has expanded its scope and mission, becoming deeply engaged in research, outreach, training, and consulting. More than 1,200 people have attended our workshops around the country, and many of them have become intensely involved in redistricting projects. We think the time is right to make a computational intervention. The mathematics of gerrymandering is surprisingly rich—enough to launch its own subfield—and computing power is arguably just catching up with the scale and complexity of the redistricting problem. Despite our group's technical orientation, our central goal is to reinforce and protect civil rights, and we are working closely with lawyers, political scientists, geographers, and community groups to build tools and ideas in advance of the next U.S. Census and the round of redistricting to follow it.

In a country that vests power in elected representatives, there will always be skirmishes for control of the electoral process. And in a system such as that of our House of Representatives—where winner takes all within each geographical district—the delineation of voting districts is a natural battleground. American history is chock-full of egregious line-drawing schemes, from stuffing a district with an incumbent's loyalists to slicing a long-standing district three ways to suppress the political power of black voters. Many varieties of these so-called *packing and cracking* strategies continue today, and in the big data moment, they have grown enormously more sophisticated. Now more than ever, abusive redistricting is stubbornly difficult to even identify definitively. People think they know gerrymandering by two hallmarks—bizarre shapes and disproportionate electoral outcomes—yet neither one is reliable. So how do we determine when the scales are unfairly tipped?

The Eyeball Test

The 1812 episode that gave us the word "gerrymander" sprang from the intuition that oddly shaped districts betray an illegitimate agenda. It is named for Elbridge Gerry, who was governor of Massachusetts at the time. Gerry had quite a Founding Father pedigree—signer of the Declaration of Independence, major player at the U.S. Constitutional Convention, member of Congress, James Madison's vice president—so it is amusing to consider that his enduring fame comes from nefarious redistricting. "Gerry-mander," or Gerry's salamander, was the satirical name given to a curvy district in Boston's North Shore that was thought to favor the governor's Democratic-Republican party over the rival Federalists. A woodcut political cartoon ran in the *Salem Gazette* in 1813; in it, wings, claws, and fangs were suggestively added to the district's contours to heighten its appearance of reptilian contortions.

So the idea that erratic districts tip us off to wrongdoing goes a long way back, and the converse notion that close-knit districts promote democratic ideals is as old as the republic. In 1787, Madison wrote in *The Federalist Papers* that "the natural limit of a democracy is that distance from the central point which will just permit the most remote citizens to assemble as often as their public functions demand." In other words, districts should be transitable. In 1901, a federal apportionment act marked the first appearance in U.S. law of the vague desideratum that districts should be composed of "compact territory." The word "compact" then proliferated throughout the legal landscape of redistricting but almost always without a definition.

For instance, at a 2017 meeting of the National Conference of State Legislatures, I learned that after the last census, Utah's lawmakers took the commendable time and effort to set up a website, Redistrict Utah, to solicit proposed districting maps from everyday citizens. To be considered, maps were required to be "reasonably compact." I jumped at the opportunity to find out how exactly that quality was being tested and enforced, only to learn that it was handled by just tossing the funny-looking maps. If that sounds bad, Utah is far from alone. Thirty-seven states have some kind of shape regulation on the books, and in almost every case, the eyeball test is king.

The problem is that the outline of a district tells a partial and often misleading story. First, there can certainly be benign reasons for ugly shapes. Physical geography or reasonable attempts to follow county lines or unite communities of interest can influence a boundary, although just as often, legitimate priorities such as these are merely scapegoated in an attempt to defend the worst-offending districts. On the other hand, districts that are plump, squat, and symmetrical offer no meaningful seal of quality. Just this year, a congressional redistricting plan in Pennsylvania drafted by Republicans in the state legislature achieved strong compactness scores under all five formulas specified by Pennsylvania's supreme court. Yet mathematical analysis revealed that the plan would nonetheless lock in the same extreme partisan skew as the contorted plan, enacted in 2011, that it was meant to replace. So the justices opted for the extraordinary measure of adopting an independent outsider's plan.

Lopsided Outcomes

If shape is not a reliable indicator of gerrymandering, what about studying the extent to which elected representatives match the voting patterns of the electorate? Surely lopsided outcomes provide prima facie evidence of abuse. But not so fast. Take Republicans in my home state of Massachusetts. In the 13 federal elections for president and Senate since 2000, GOP candidates have averaged more than one third of the votes statewide. That is six times the level needed to win a seat in one

The Power of the Pen

Gerrymandering relies on carefully drawn lines that dilute the voting power of one population to favor another by clustering one side's voters into a few districts with excessively high numbers (packing), by dispersing them across several districts so that they fall short of electing a preferred candidate (cracking), or by using a combination of the two schemes.

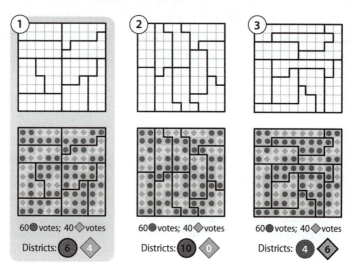

60 ● votes; 40 ◇ votes   60 ● votes; 40 ◇ votes   60 ● votes; 40 ◇ votes

Districts: (6) ◇4   Districts: (10) ◇0   Districts: (4) ◇6

A grid is districted to produce an electoral outcome proportional to the share of votes for each party **1**. The same grid can be districted using combinations of packing and cracking to produce extreme outcomes **2, 3**—one in which the Blue party wins all districts and one in which it wins only four of 10. In this particular case, the geometry of the layout turns out to favor the Blue party. Statistical analysis using Markov chain Monte Carlo reveals that the Orange party is far more likely to get two or three seats, rather than its proportional share of four, in the universe of possible plans. See also color images.

of Massachusetts's nine congressional districts because a candidate in a two-way race needs a simple majority to win. Yet no Republican has won a seat in the House since 1994.

We must be looking at a gerrymander that denies Republicans their rightful opportunity districts, right? Except the mathematics here is completely exonerating. Let us look at a statewide race so that we can put uncontested seats and other confounding variables to the side. Take Kenneth Chase, the Republican challenger to Ted Kennedy for the U.S. Senate in 2006, who cracked 30% of the statewide vote. Proportionally, you would expect Chase to beat Kennedy in nearly three out of nine congressional districts. But the numbers do not shake out. As it turns out, it is mathematically impossible to select a single district-sized grouping of towns or precincts, even scattered around the state, that preferred Chase. His voters simply were not clustered enough. Instead, most precincts went for Chase at levels close to the state average, so there were too few Chase-favoring building blocks to go around.

Any voting minority needs a certain level of nonuniformity in how its votes are distributed for our districting system to offer even a theoretical opportunity to secure representation. And the type of analysis applied to the Chase–Kennedy race does not even consider spatial factors, such as the standard requirement that each district be one connected piece. One may rightfully wonder how we can ever hold district architects accountable when the landscape of possibilities can hold so many surprises.

Random Walks to the Rescue

The only reasonable way to assess the fairness of a districting plan is to compare it with other valid plans for cutting up the same jurisdiction because you must control for aspects of electoral outcomes that were forced by the state's laws, demographics, and geography. The catch is that studying the universe of possible plans becomes an intractably big problem.

Think of a simple four-by-four grid and suppose that you want to divide it into four contiguous districts of equal size, with four squares each. If we imagine the grid as part of a chessboard, and we interpret contiguity to mean that a rook should be able to visit the entire district, then there are exactly 117 ways to do it. If corner adjacency is permitted—so-called queen contiguity—then there are 2,620 ways. And they are not so straightforward to count. As my colleague Jim Propp, a professor at the University of Massachusetts Lowell and a leader in the field of combinatorial enumeration, puts it, "In one dimension, you can split paths along the way to divide and conquer, but in two dimensions, suddenly there are many, many ways to get from point A to point B."

The issue is that the best counting techniques often rely on recursion—that is, solving a problem using a similar problem that is a step smaller—but two-dimensional spatial counting problems just do not recurse well without some extra structure. So complete enumerations must rely on brute force. Whereas a cleverly programmed laptop can classify partitions of small grids nearly instantly, we see huge jumps in complexity as the grid size grows, and the task quickly zooms out of reach. By the time you get to a grid of nine-by-nine, there are more than 700 trillion solutions for equinumerous rook partitions, and even a high-performance computer needs a week to count them all. This seems like a hopeless state of affairs. We are trying to assess one way of cutting up a state without any ability to enumerate—let alone meaningfully compare it against—the universe of alternatives. This situation sounds like groping around in a dark, infinite wilderness.

The good news is that there is an industry standard used across scientific domains for just such a colossal task: Markov chain Monte Carlo (MCMC). Markov chains are random walks in which where you go next is governed by probability, depending only on where you are now (at every position, you roll the dice to choose a neighboring space to move to). Monte Carlo methods are just estimation by random sampling. Put them together, and you get a powerful tool for searching vast spaces of possibilities. MCMC has been successfully used to decode prison messages, probe the properties and phase transitions of liquids, find provably accurate fast approximations for hard computational problems, and much more. A 2009 survey by the eminent statistician

---

How to Compare Countless Districting Plans

Markov chains are random walks around a graph or network in which the next destination is determined by a probability, like a roll of the dice, depending on the current position. Monte Carlo methods use random sampling to estimate a distribution of probabilities. Combined, Markov chain Monte Carlo (MCMC) is a powerful tool for searching and sampling from a vast space of scenarios, such as all the possible districting plans in a state. Attempts to use computational analysis to spot devious districting go back several decades, but efforts to apply MCMC to the problem are much more recent.

Dimensions; Districts	Equal-Size Districts	District Sizes Can Be Unequal (±1)
2×2 grid; 2 districts	2	6
3×3 grid; 3 districts	10	58
4×4 grid; 2 districts	70	206
4×4 grid; 4 districts	117	1,953
4×4 grid; 8 districts	36	34,524
5×5 grid; 5 districts	4,006	193,152
6×6 grid; 2 districts	80,518	?*
6×6 grid; 3 districts	264,500	?
6×6 grid; 4 districts	442,791	?
6×6 grid; 6 districts	451,206	?
6×6 grid; 9 districts	128,939	?
6×6 grid; 12 districts	80,092	?
6×6 grid; 18 districts	6,728	?
7×7 grid; 7 districts	158,753,814	?
8×8 grid; 8 districts	187,497,290,034	?
9×9 grid; 9 districts	706,152,947,468,301	?

*Mathematicians have not yet enumerated these solutions, which can require a week of computing or more. To find out more about the hunt for these numbers, visit www.mggg.org.

Equal-size districts: 2 solutions

District size can be +/− 1: 6 solutions

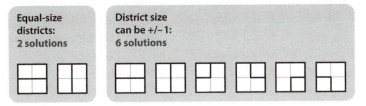

SIMPLE CASE. It is easy to enumerate all the ways to partition a small grid into equal-size districts. For a two-by-two grid with two districts of equal size, there are only two solutions. But if districts can vary in size, the number of solutions jumps to six.

Persi Diaconis estimated that MCMC drives 10 to 15% of the statistical work in science, engineering, and business, and the number has probably only gone up since then. Although computational analysis in redistricting goes back several decades, serious attempts to apply MCMC in that effort only started to appear publicly around 2014.

Imagine that officials in the state of Gridlandia hire you to decide if their legislature's districting plan is reasonable. If Gridlandia is a four-by-four grid of squares, and its state constitution calls for rook-contiguous districts, then you are in luck: There are exactly 117 ways to produce a compliant plan, and you can examine them all. You can set up a perfectly faithful model of this universe of districting plans by using 117 nodes to represent the valid plans and adding edges between the nodes to represent simple moves in which two squares in the grid swap their district assignments. The edges give you a way of conceptualizing how similar two plans are by simply counting the number of swaps needed to transform one to the other. (I call this structure a *metagraph* because it is a graph of ways to cut up another graph.) Now suppose that the state legislature is controlled by the Diamond party, and its rivals suspect that it has rigged the seats in its favor. To determine if that is true, one may turn to the election data. If the Diamond plan would have produced more seats for the party in the last election than, say, 114 out of 117 alternatives and if the same is true for several previous elections, the plan is clearly a statistical outlier. This is persuasive evidence of a partisan gerrymander—and you do not need MCMC for such an analysis.

The MCMC method kicks in when you have a full-sized problem in place of this small toy problem. As soon as you get past 100 or so nodes, there is a similar metagraph, but you cannot completely build it because of its forbidding complexity. That is no deal breaker, though. From any single plan, it is still easy to build out the local neighborhood by performing all possible moves. Now you can take a million, billion, or trillion steps and see what you find. There is mathematics in the background (ergodic theory, to be precise) guaranteeing that if you random-walk for long enough, the ensemble of maps you collect will have properties representative of the overall universe, typically long before you have visited even a modest fraction of nodes in your state space. This procedure lets you determine if the map you are evaluating is an extreme outlier according to various partisan metrics.

The cutting edge of scientific inquiry is to build more powerful algorithms and, at the same time, to devise new theorems that certify that we are sampling well enough to draw robust conclusions. There is an emerging scientific consensus around this method but there are also many directions of ongoing research.

RIP Governor Gerry

So far, courts seem to be smiling on this approach. Two mathematicians—Duke University's Jonathan Mattingly and Carnegie Mellon University's Wes Pegden—have recently testified about MCMC approaches for the federal case in North Carolina and the state-level case in Pennsylvania, respectively.

Mattingly used MCMC to characterize the reasonable range one might observe for various metrics, such as seats won, across ensembles of districting plans. His random walk was weighted to favor plans that were deemed closer to ideal, along the lines of North Carolina state law. Using his ensembles, he argued that the enacted plan was an extreme partisan outlier. Pegden used a different kind of test, appealing to a rigorous theorem that quantifies how unlikely it is that a neutral plan would score much worse than other plans visited by a random walk. His method produces p-values, which constrain how improbable it is to find such anomalous bias by chance. Judges found both arguments credible and cited them favorably in their respective decisions.

For my part, Pennsylvania governor Tom Wolf brought me on earlier this year as a consulting expert for the state's scramble to draw new district lines following its supreme court's decision to strike down the 2011 Republican plan. My contribution was to use the MCMC framework to evaluate new plans as they were proposed, harnessing the power of statistical outliers while adding new ways to take into account more of the varied districting principles in play, from compactness, to county splits, to community structure. My analysis agreed with Pegden's in flagging the 2011 plan as an extreme partisan outlier—and I found the new plan floated by the legislature to be just as extreme, in a way that was not explained away by its improved appearances.

As the 2020 Census approaches, the nation is bracing for another wild round of redistricting, with the promise of litigation to follow. I hope the next steps will play out not just in the courtrooms but also in reform measures that require a big ensemble of maps made with open source tools to be examined before any plan is signed into law. In that way, the legislatures preserve their traditional prerogatives to commission and approve district boundaries, but they have to produce some guarantees that they are not putting too meaty a thumb on the scale.

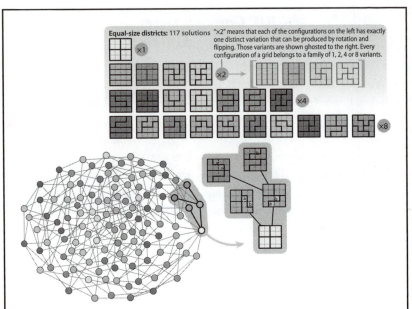

BIGGER CASE. As the size of the grid grows, the number of possibilities for carving it up skyrockets. Dividing a four-by-four grid into four districts of equal size has 117 solutions. If the districts can vary in size by even one unit, there are 1,953 solutions. It does not take long before even the most powerful computers struggle to enumerate the possibilities for more complex grids. That situation presents a problem for anyone trying to detect manipulative maps by comparing the myriad ways to district a U.S. state. But MCMC can help.

We can efficiently explore valid districting plans by traveling randomly around a metagraph, defined by moves such as the unit swaps pictured. In the highlighted inset, each pattern has squares marked **a** and **b** whose district assignments are exchanged to arrive at the configuration of the pattern shown. The edges in the network represent these simple swap moves. The metagraph models the space of all valid districting plans and can be used to sample many billions of plans. Geometers are trying to understand the shape and structure of that universe. See also color images.

Computing will never make tough redistricting decisions for us and cannot produce an optimally fair plan. But it can certify that a plan behaves as though selected just from the stated rules. That alone can rein in the worst abuses and start to restore trust in the system.

More to Explore

A Formula Goes to Court: Partisan Gerrymandering and the Efficiency Gap. Mira Bernstein and Moon Duchin in *Notices of the American Mathematical Society*, Vol. 64, No. 9, pp. 1020–1024; October 2017. www.ams.org/journals/notices/201709/rnoti -p1020.pdf.

Gerrymandering Metrics: How to Measure? What's the Baseline? Moon Duchin in *Bulletin of the American Academy of Arts & Sciences*, Vol. 71, No. 2, pp. 54–58; Winter 2018.

Slicing Sandwiches, States, and Solar Systems: Can Mathematical Tools Help Determine What Divisions Are Provably Fair?

Theodore P. Hill

Gerrymandering is making headlines once again, with a case already before the Supreme Court regarding partisan redistricting in Wisconsin and another from Pennsylvania waiting in the wings. At the core of the problem of redrawing congressional districts is the issue of fairness, and that is tricky business indeed. The general subject of fair division has been studied extensively using mathematical tools, and some of that study has proved very useful in practice for problems such as dividing estates or fishing grounds. For gerrymandering, however, there is still no widely accepted fair solution. On the contrary, this past October, Pablo Soberón of Northeastern University showed that a biased cartographer could apply mathematics to gerrymander on purpose, without even using strange shapes for the districts. The underlying idea traces back to one of mathematicians' favorite theorems, which dates back to World War II.

The late 1930s were devastating years for the Polish people, but they were years of astonishing discovery for Polish mathematicians. Between the rock of the Great Depression and the hard place of impending invasion and occupation by both Nazi and Soviet armies, a small group of mathematicians from the university in Lwów (today Lviv) met regularly in a coffee shop called the Scottish Café to exchange mathematical ideas. These ideas were not the mathematics of complicated calculations (which were then done with the aid of slide rules) but rather were very general and aesthetically beautiful abstract concepts, soon to prove extremely powerful in a wide variety of mathematical and scientific fields.

The café tables had marble tops and could easily be written on in pencil and then later erased like a slate blackboard. Since the group often returned to ideas from previous meetings, they soon realized the need for a written record of their results and purchased a large notebook for documenting the problems and answers. The book, kept in a safe place by the café headwaiter and produced by him upon the group's next visit, was a collection of these mathematical questions, both solved and unsolved, that decades later became known in international mathematical circles as the *Scottish Book*.

The Ham Sandwich Problem

Problem No. 123 in the book, posted by Hugo Steinhaus, a senior member of the café mathematics group and a professor of mathematics at the University of Lemberg (now the University of Lviv), was stated as follows:

> Given are three sets A_1, A_2, A_3, located in the three-dimensional Euclidean space and with finite Lebesgue measure. Does there exist a plane cutting each of the three sets A_1, A_2, A_3, into two parts of equal measure?

To bring this question to life for his companions, Steinhaus illustrated it with one of his trademark vivid examples, one that reflected the venue of their meetings, and also perhaps their imminent preoccupation with daily essentials: Can every ordinary ham sandwich consisting of three ingredients, say bread, ham, and cheese, be cut by a planar slice of a knife so that each of the three is cut exactly in half?

A Simpler Problem

At the meeting where Steinhaus introduced this question, he reported that the analogous conclusion in two dimensions was true: Any two areas in a (flat) plane can always be simultaneously bisected by a single straight line, and he sketched out a solution on the marble tabletop. In the spirit of Steinhaus's food theme, let's consider the case where the two areas to be bisected are the crust and sausage on a pepperoni pizza. If the pizza happens to be a perfect circle, then every line passing through its center exactly bisects the crust.

To see that there is always a line that bisects both crust and sausage simultaneously, start with the potential cutting line in any fixed direction and rotate it about the center slowly, say, clockwise. If the proportion of sausage on the clockwise side of the arrow-cut happened to be 40% when the rotation began, then after the arrow-cut has rotated 180 degrees, the proportion on the clockwise side of the arrow-cut is now 60%. Because this proportion changed continuously from 40% to 60%, at some point it must have been exactly 50%, and at that point both crust and sausage have been exactly bisected (Figure 1).

On the other hand, if the pizza is not a perfect circle, as no real pizza is, then there may not be an exact center point such that every straight line through it exactly bisects the crust. But in this general noncircular case, again move the cutting line so that it always bisects the crust as it rotates, and note that even though the cutting line may not rotate around a single point as it did with a circular pizza, the same continuity

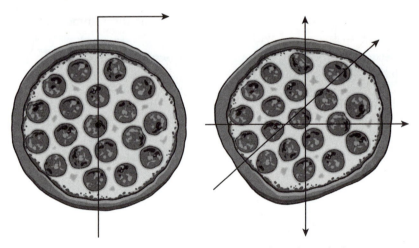

FIGURE 1. If a pizza is a perfect circle, then every line through the center bisects the crust. If the cut starts with 40% of the sausage clockwise from the arrow, after rotating 180 degrees, 60% of the sausage is clockwise from the arrow. So somewhere in between, the line hits 50% and the same cutting line bisects both crust and sausage. If the pizza is not a perfect circle, the crust-bisecting lines may not all pass through the same point, but the same argument applies.

argument applies. If the proportion clockwise of the north cut started at 40%, then when the cut arrow points south, that proportion will be 60%, which again completes the argument using the simple fact that to go continuously from 40 to 60, one must pass through 50. This simple but powerful observation, formally known as the intermediate value theorem, also explains why if the temperature outside your front door was 40 degrees Fahrenheit yesterday at noon and 60 degrees today at noon, then at some time in between, perhaps several times, the temperature must have been exactly 50 degrees.

Steinhaus's two-dimensional (pizza) version of the ham sandwich theorem may be used for gerrymandering. Instead of a pizza, imagine a country with two political parties whose voters are sprinkled through it in any arbitrary way. The pizza theorem implies that there is a straight line bisecting the country so that exactly half of each party is on each side of the line. Suppose, for example, that 60% of the voters in the United States are from party Purple and 40% are from party Yellow. Then there is a single straight line dividing the country into two regions, each of which has exactly 30% of the Purple on each side, and exactly 20% of the Yellow on each side, so the Purple have the strict majority on both sides. Repeating this procedure to each side yields four districts with exactly 15% Purple and exactly 10% Yellow in each. Again the majority party (in this case, Purple) has the majority in each district. Continuing this argument shows that whenever the number of desired districts is a power of two, there is always a straight-line partition of the country into that number of districts so that the majority party also has the majority of votes in every single district (Figure 2).

This repeated-bisection argument may fail, however, for odd numbers of desired districts. On the other hand, Sergei Bespamyatnikh, David Kirkpatrick, and Jack Snoeyink of the University of British Columbia found a generalization of the ham sandwich theorem that does the trick for any number of districts, power of two or not. They showed that for a given number of Yellow and Purple points in the plane (no three of which are on a line), there is always a subdivision of the plane into any given number of convex polygons (districts), each containing exactly the same numbers of Yellow points in each district, and the same number of Purple (Figure 3).

In his application of this theorem to gerrymandering, Soberón observed that for any desired number of districts, this theorem implies

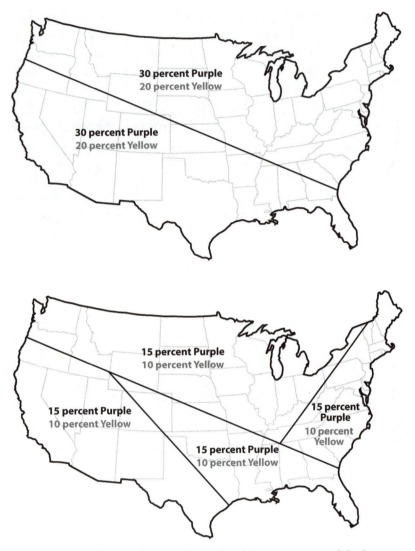

FIGURE 2. According to the two-dimensional (pizza) version of the ham sandwich theorem, there is a straight line across the United States so that exactly half of the Purple and half of the Yellow party voters are on either side (top). Bisecting each of those (bottom), the same argument shows that there are four regions with equal numbers of Purple and equal numbers of Yellow in each of them. Thus, the party with the overall majority also has the majority in each of the districts. See also color images.

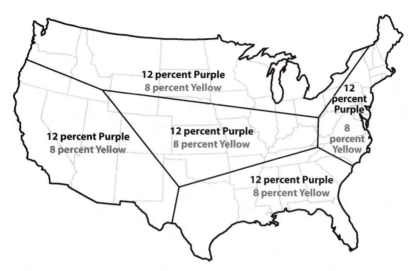

FIGURE 3. For odd numbers of desired districts, the repeated-bisection argument of the two-dimensional version of the ham sandwich theorem may fail. However, a generalization of the theorem works for any number of districts, by showing that for a given number of Purple or Yellow points in a plane (no three of which are on a line), there is always a subdivision of the plane into any given number of convex polygons, each of which contains exactly the same number of Yellow, and the same number of Purple, points. See also color images.

that there is always a subdivision into that number of polygonal districts so that each district has exactly the same number of Purple and exactly the same number of Yellow. Whichever party has the overall majority in the country also has the majority in every district. Thus, as he found, a direct application of the ham sandwich theory would not help fix the problem, but would actually make it worse, and the electorate should be wary if the person drawing congressional maps knows anything about that theory. No wonder the Supreme Court balked on all three of the most recent cases it has heard on partisan gerrymandering.

The Scottish Café

After giving his argument for the two-dimensional case of the ham sandwich theorem, Steinhaus then challenged his companions to prove the three-dimensional version. The same basic intermediate value theorem

argument of continuity that worked for the pizza theorem does not settle the "ham sandwich" Problem 123 question, simply because there is
no single "direction" to move a given starting plane passing through the
sandwich, guaranteeing a return to the same spot having bisected both
of two other objects somewhere along the way.

Two gifted students and protégés of Steinhaus, Stefan Banach and
Stanisław Ulam, were also members of the Scottish Café group. Using a
discovery Ulam had made around the same time with Karol Borsuk, another Scottish Café comrade, Banach was able to prove the sandwich conjecture of Steinhaus. The key to Banach's proof, called the Borsuk–Ulam
theorem, was another general continuity theorem similar in spirit to the
intermediate value theorem but much more sophisticated. Steinhaus also
brought that abstract theorem to life with another of his colorful real-life
examples: the Borsuk–Ulam theorem, he said, implies that at any given
moment in time there are two antipodal points on the Earth's surface that
have the same temperature and the same atmospheric pressure.

If there are more than three solid objects, or more than two regions in
the plane, then it may not be possible to bisect all of them simultaneously
with a single plane (or line), as can easily be seen in the case where four
small balls are located at the vertices of a pyramid. Also the conclusion of
bisection cannot generally be relaxed. For example, if your goal is to split
a pizza (or political territory) into two pieces so that one side contains
exactly 60% of each, that may not always be possible (Figure 4).

Generalizations

During World War II, the statement of this colorful and elegant new
mathematical result—that any three fixed objects simultaneously can
be bisected by a single plane—somehow made it through enemy territory and across the Atlantic, long before e-mail or smart phones or
Skype. Mathematicians Arthur Stone and John Tukey at Princeton
University learned about this new gem of a theorem via the international mathematics grapevine and improved the result to include
nonuniform distributions, higher dimensions, and a variety of other
cutting surfaces and objects. The new Stone and Tukey extensions also
showed, for example, that a single circle simultaneously can bisect any
three shapes in the plane. For example, there is a location for a telecommunications satellite and a power level so that its broadcasts will

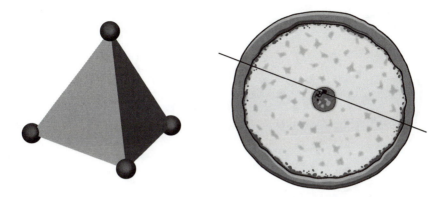

FIGURE 4. It is not always possible to bisect simultaneously more than three objects with a single plane (such as points at the corners of a pyramid, shown at left), nor to separate simultaneously three objects by the same unequal ratios. The analog in two dimensions (right) shows that the pizza cannot be cut by a straight line so that exactly 60% of the crust and 60% of the sausage are on the same side of the line.

reach exactly half the Yellow, half the Purple, and half the Teal (Independents) (Figure 5).

Formally speaking, of course, drawing a line to bisect two discrete mass distributions such as Yellow and Purple voters may require splitting one of the voter points, which may not always be possible (or desirable). If a distribution has an odd number of indivisible points of one type, for example, then clearly no line can have exactly half those points on each side of the line. Inspired by the success of my Ph.D. advisor, Lester Dubins, in addressing a different fair division problem involving indivisible points (professors, in that case), I wondered whether the conclusion of the ham sandwich theorem might be extended to also include mass distributions with indivisible points—such as grains of salt and pepper sprinkled on a table—by replacing the notion of exact bisection of distributions by a natural generalization of the statistical notion of a median.

Recall that a median of a distribution, say of house prices in a neighborhood, is a price such that no more than half of all the house values are below and no more than half are above that price. Extending this notion to higher dimensions yields the concept of median lines,

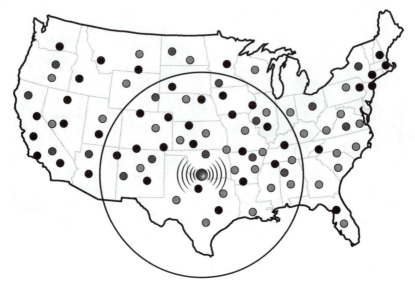

FIGURE 5. Mathematicians Arthur Stone and John Tukey of Princeton University extended the ham sandwich theorem to nonuniform distributions, higher dimensions, and a variety of other cutting surfaces and objects. One of their examples showed that a single circle simultaneously can bisect any three shapes in the plane. For instance, it is always possible to design the power and location of a telecommunications satellite so that its broadcasts reach exactly half the Yellow, half the Purple, and half the Teal (Independents). See also color images.

median planes, and median hyperplanes in higher dimensions. Using the Borsuk–Ulam theorem again, but this time applied to a different "midpoint median" function, it was straightforward to show that for any two arbitrary random distributions in the plane, or any three in space, there is always a line median or plane median, respectively, that has no more than half of each distribution on each side.

Some 20 years later, Columbia University economist Macartan Humphreys used this result to solve a problem in cooperative game theory. In a setting where several groups must agree on allocations of a fixed resource (say, how much of a given disaster fund should be allocated to medical, power, housing, and food), the objective is to find an allocation that no winning coalition could override in favor of another allocation. He showed that such equilibrium allocations exist precisely when they lie on "ham sandwich cuts."

Touching Planes

In explaining the beauties of the ham sandwich theorem to nonmathematician friends over beer and pizza, one of my companions noticed that often there is more than one bisecting line (or plane), and we saw that some bisecting lines might touch each of the objects, whereas others may not. I started looking at this observation more closely and discovered that in every case, I could always find a bisecting line or plane that touched all the objects. When I could not find a reference or proof of this concept, I posed the question to my Georgia Tech friend and colleague John Elton, who had helped me crack a handful of other mathematical problems: Is there always a bisecting plane (or hyperplane, in dimensions greater than 3) that also touches each of the objects?

Together, he and I were able to show that the answer is yes, which strengthens the conclusion of the classical ham sandwich theorem. For example, this improved version implies that at any instant in time in our solar system, there is always a single plane passing through three bodies—one planet, one moon, and one asteroid—that simultaneously bisects the planetary, the lunar, and the asteroidal masses in the solar system (Figure 6).

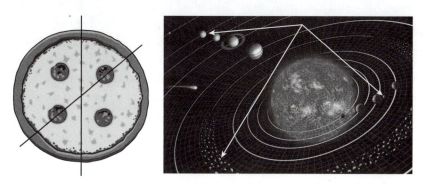

FIGURE 6. Some bisecting lines or planes may touch each of the objects, whereas others may not, as shown on the pizza above. Nevertheless, there is always a single bisecting line or plane (or hyperplane, in higher dimensions) that touches all of the objects. For example, at any instant in time in our solar system, there is always a single plane passing through three bodies— one planet, one moon, and one asteroid—that simultaneously bisects the planetary, the lunar, and the asteroidal masses in the solar system.

Diverse Divisions

The ideas underlying the ham sandwich theorem have also been used in diverse fields, including computer science, economics, political science, and game theory. When I asked my friend Francis Su, Harvey Mudd College mathematician and fair-division expert, about his own applications of the ham sandwich theorem, he explained how he and Forest Simmons of Portland Community College had used ham sandwich results to solve problems in *consensus halving*. In particular, they used it to show that given a territory and $2n$ explorers, two each of n different specialties (e.g., two zoologists, two botanists, and two archaeologists), there always exists a way to divide the territory into two regions and the people into two teams of n explorers (one of each type) such that each explorer is satisfied with their half of the territory.

As a more light-hearted application during a keynote lecture at Georgia Tech, Tel Aviv University mathematician Noga Alon described a discrete analog of the ham sandwich theorem for splitting a necklace containing various types of jewels, as might be done, he said, by mathematically oriented thieves who steal a necklace and wish to divide it fairly between them. Even though it had been offered as an amusement, his result had applications, including to very large scale integrated (VLSI) circuit designs where an integrated chip composed of two different types of nodes is manufactured in the shape of a closed circuit (much like a necklace), and may be restructured after fabrication by cutting and regrouping the pieces. Alon's theorem answers this question: How many cuts need to be made of the original circuit in order to bisect it into two parts, each containing exactly half of each type of node?

Revisiting the Café

Steinhaus published the proof of the ham sandwich theorem in the local Polish mathematical journal *Mathesis Polska* in 1938, the year of the infamously violent *Kristallnacht*. The Scottish Café mathematics gatherings continued for a few more years, despite the invasion of western Poland by the German army and the Soviet occupation of Lwów from the east, but the difficult times would soon disperse both scholars and their works. Ulam, a young man in his 20s and, like Steinhaus, also of

Jewish roots, had left with his brother on a ship for America just two weeks before the German invasion.

Banach, nearing 50 and already widely known for his discoveries in mathematics, was appointed dean of the University of Lwów's department of mathematics and physics by the Soviets after they occupied that city, under the condition that he promised to learn Ukrainian. When the Nazis in turn occupied Lwów, they closed the universities, and Banach was forced to work feeding lice at a typhus research center, which at least protected him from being sent into slave labor. (Banach, like many others, was made to wear cages of lice on his body, so that they could feed on his blood. The lice, which are carriers of typhus, were used in research efforts to create a vaccine against the disease.) Banach was able to help reestablish the university after Lwów was recaptured by the Soviets in 1944, but he died of lung cancer in 1945.

Although the correct statement of the crisp ham sandwich theorem had made it through the World War II mathematical grapevine perfectly, the proper credit for its discoverers was garbled en route, and Stone and Tukey mistakenly attributed the first proof to Ulam. Sixty years later, the record was set straight when a copy of Steinhaus's article in *Mathesis Polska* was finally tracked down, and we now know that Steinhaus posed the problem and published the first paper on it, but it was Banach who actually solved it first, using a theorem of Ulam's.

Today Banach is widely recognized as one of the most important and influential mathematicians of the twentieth century, and many fundamental theorems, as well as entire basic fields of mathematics, that are based on his work are now among the most extensively used tools in physics and mathematics.

Ulam went on to work as one of the key scientists on the Manhattan Project in Los Alamos, New Mexico, achieving fame in particular for the Teller–Ulam thermonuclear bomb design and for his invention of Monte Carlo simulation, a ubiquitous tool in economics, physics, mathematics, and many other areas of science, which is used to estimate intractable probabilities by averaging the results of huge numbers of computer simulations of an experiment.

After the war, Steinhaus would have been welcomed with a professorship at almost any university in the world, but he chose to stay in Poland to help rebuild Polish mathematics, especially at the university in Wrocław, which had been destroyed during the war. During those years in hiding, Steinhaus had also been breaking ground on the mathematics of fair division—the study of how to partition and allocate portions of a single heterogeneous commodity, such as a cake or piece of land, among several people with possibly different values. One of Steinhaus's key legacies was his insight to take the common vague concept of "fairness" and put it in a natural and concrete mathematical framework. From there, it could be analyzed logically, and it has now evolved into common and powerful tools. For example, both the website Spliddit, which provides free mathematical solutions to complicated everyday fair division problems from sharing rent to dividing estates, and the eBay auction system, which determines how much you pay—often below your maximum bid—are direct descendants of Steinhaus's insights on how to cut a cake fairly.

These ideas, born of a mathematician living and working clandestinely with little contact with the outside world for long periods of time and undoubtedly facing fair-allocation challenges almost daily, have inspired hundreds of research articles in fields from computer and political science to theoretical mathematics and physics, including many of my own. Steinhaus eventually became the first dean of the department of mathematics in the Technical University of Wrocław. Although I never met him in person, I had the good fortune to be invited to visit that university in December 2000, and it was my privilege to lodge in a special tower suite right above the mathematics department and to give a lecture in the Hugo Steinhaus Center.

Steinhaus had made the last entry in the original *Scottish Book* in 1941, just before he went into hiding with a Polish farm family, using the assumed name and papers of a deceased forest ranger. The *Scottish Book* itself also disappeared then, and when he came out of hiding and was able to rediscover the book, Steinhaus sent a typed version of it in Polish to Ulam at Los Alamos, who translated it into English. Mathematician R. Daniel Mauldin at the University of North Texas, a friend of Ulam, published a more complete version of the *Scottish Book*, including comments and notes by many of the problems' original authors. Their Problem 123, which evolved into the ham sandwich theorem, continues to fascinate and inspire researchers, and Google Scholar shows that eight decades later, several dozen new entries on the topic still appear every few months.

But what about that pesky gerrymandering problem? Negative results in science can also be very valuable; they can illuminate how a certain line of reasoning is doomed to failure and inspire searches in other directions. That outcome is exactly what happened when the negative ham sandwich gerrymandering result showed that a redisricting attempt might still be radically biased even if the shapes of the districts are quite regular. That insight led researchers to drop the notion of shape as the key criterion and to look for another approach. The result was a new "efficiency gap" formula that quantifies how much a map is gerrymandered based on vote shares, not on shape. This formula, too, has problems, and in turn it inspired me and my colleague Christian Houdré at Georgia Tech to look for a better measure of "gerrymandered-ness" using combinatorial models involving balls and urns. And so the exciting cycle of scientific discovery that started with the ham sandwich theorem continues.

A great many mathematicians today owe a huge debt to those intrepid Polish academics, and we raise our cups of java to those original Scottish Café mathematicians (Figure 7)!

 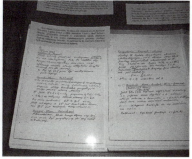

FIGURE 7. The building that housed the Scottish Café where the group of mathematicians met in the late 1930s is still standing in Lwów (left, photo by Stanisław Kosiedowski). Copies of the *Scottish Book,* with the original entries by Banach and Ulam are on display at the Library of the Mathematical Institute of the Polish Academy of Sciences in Warsaw (right, photo from PIWiki, uploaded by Stako). The original book remains in the custody of the Banach family, which took it with them after Banach's death and the war ended, when they were required to resettle in Warsaw. Steinhaus kept in touch with the family, and after the war, copied the book by hand to send to Ulam at Los Alamos in 1956. Ulam translated the book into English and had 300 copies made at his own expense. Requests for the book became so numerous that another edition was printed in 1977. After a *Scottish Book* Conference in 1979, in which Ulam participated, the book was again reissued with updated material and additional papers.

Bibliography

Bellos, A. 2014. *The Grapes of Math*. New York: Simon and Schuster.

Bespamyatnikh, S., D. Kirkpatrick, and J. Snoeyink, J. Generalizing ham sandwich cuts to equitable subdivisions. *Discrete and Computational Geometry* 24: 605–622.

Elton, J., and T. Hill. 2011. A stronger conclusion to the classical ham sandwich theorem. *European Journal of Combinatorics* 32: 657–661.

Hill, T. 2000. Mathematical devices for getting a fair share. *American Scientist* 88: 325–331.

Humphreys, M. 2008. Existence of a multicameral core. *Social Choice and Welfare* 31: 503–520.

Mauldin, R. D. (ed.) 2015. *The Scottish Book: Mathematics from the Scottish Cafe*, 2nd ed. Basel, Switzerland: Birkhäuser.

Simmons, F., and F. Su. 2003. Consensus-halving via theorems of Borsuk-Ulam and Tucker. *Mathematical Social Sciences* 45: 15–25.

Soberón, P. 2017. Gerrymandering, Sandwiches, and Topology. *Notices of the American Mathematical Society* 64: 1010–1013.

Steinhaus, H. 1938. A note on the ham sandwich theorem. *Mathesis Polska* XI: 26–28.

Does Mathematics Teach How to Think?

Paul J. Campbell

What are the larger benefits of learning mathematics? We are not referring to what is variously termed number sense, numeracy, quantitative literacy, or quantitative reasoning.

In a tradition that goes back to Plato in his *Republic*, educators have maintained that mathematics beyond arithmetic is an essential component of an education: It "trains the mind," by teaching logical thinking and abstraction.

> *Mathematics . . . teaches . . . how to think . . . .* Reasoning is learned by practice and there is no better practice than mathematics. We have problems that can be solved by reasoning and we can see that our reasoning leads to correct answers. This is an advantage over any other subject . . . .
>
> [Dudley 2008, 2]

Hence, in an earlier era, students studied geometry from Euclid, memorizing and understanding proofs of theorems about idealized geometrical objects.

What habits of mind are ascribed to the learning of mathematics? Some are broadly based: identifying significant information; attention to appropriate detail and exactness; inculcating the discipline of committing to memory definitions, terminology, important facts, and frequently used techniques; following patterns and systematically applying rules; and constructing and evaluating logical arguments.

Others are more specific to mathematics: deduction and argument specifically from precise definitions and principles; discerning patterns from concrete examples; a spirit of generalizing; abstraction to remove less relevant detail; searching for justifying and/or falsifying arguments about conjectures; translation among domains (words, symbols, figures); reasoning with symbols; algorithmic approaches to calculation; investigation of exceptions, boundary situations, and limiting cases; and emphasis on numerical accuracy with suitable approximation.

We examine the claims about mathematics teaching how to think, considering the merits of some alternatives to mathematics, and then ask how mathematics *as taught* can realize those goals.

Alternatives to Mathematics

What about Latin? or Classical Greek? or Even Puzzles?

Claims of training the mind were traditionally made for learning Latin (and Greek) as part of a "classical" education. Some emphasis derived from the practicality of Latin as the medium of scholarly communication in the Middle Ages, and some from the desirability for clergy to read scripture in those languages. Today—in Germany, for example—Latin is still seen in applied terms, as a prerequisite for students aspiring to careers in law or medicine.

The study of any language involves categorization of parts of speech, declension of nouns and conjugation of verbs, memorization of vocabulary and terminology, learning numerous syntax rules and patterns (and their exceptions), and developing precision in expression—not to mention the language's oral component and translation between the oral and written components. Thus, language study involves and develops many of the broad-based habits of mind delineated above. Study of an accompanying culture has other humanistic values, which we do not consider here.

The claims for both mathematics and classical languages partake of the Theory of Formal Discipline (TFD), which asserts that certain fields of study develop general mental faculties, such as observation, attentiveness, discrimination, and reasoning [Aleven 2012], and these faculties have general applicability and transfer to other domains. TFD formalizes the hunches of Plato and similar notions of Locke.

Burger [2019] makes a case for puzzle solving to teach thinking. Part of the rationale is that because solving imaginative puzzles is scarcely a field of study and has no obvious applications, it is easier to focus on the thinking processes and their development.

WHY NOT CODING?

Since the purpose of mathematics education is to improve the mind, *it does not matter much what mathematics is taught* . . . . [W]e must teach them something, so we teach them to follow rules to solve problems.

[Dudley 2008, 2–3]

If the subject matter in which one develops and practices habits of mind does not matter, why not Latin? or chess? or coding, the writing of instructions for computers? "Coding for all!" meaning that all students should learn to program computers, has become a meme of contemporary U.S. culture. Why coding? Several rationales present themselves.

Coding as Social Welfare: Jobs in information technology (IT) pay well, hence coding could provide a socioeconomic "escalator" for upward mobility of students from disadvantaged backgrounds.

Coding as Career Insurance: In the wake of the Great Recession, STEM (science, technology, engineering, mathematics) jobs provide the living standard that Americans expect, and most such jobs are computer-related.

Coding as Educational Fad: Coding is the new literacy.

Coding as Competition: Since other countries have adopted "coding for all" (e.g., the United Kingdom in 2014), the United States needs to do so too, in order to be competitive in world markets of labor, technology, and commerce.

There are potential benefits to learning coding. Coding is an attractive opportunity that can enthuse some students. Coding for all fits well with the prevailing American business model for education as primarily job training: "Computer skills," like "math skills," are valued job skills.

Relevant to habits of mind, coding involves basic logic (e.g., Boolean logic and conditionals), demands attention to syntactical detail, and involves planning ahead—which, if not one of the habits of mind above, is a valuable trait in life.

CODING: DRAWBACKS AND PRACTICAL CONSIDERATIONS

Do we need so many coders, when the demand for computer programmers is projected to decrease 8% by 2024 [Galvy 2016]? Would a career in coding be meaningful to many students? Can all students succeed at coding, or would it become just another resented obstacle (much as mathematics is now)? How low a standard would any measure of success have to meet? Would achievement gaps widen, with—just as with mathematics—some students considered "born to it" and others deemed hopeless? What other educational opportunities would instruction in coding replace? In most states, "computer science" can replace mathematics or science as a high school graduation requirement [Code.org 2016]; at some colleges, computer programming can satisfy a requirement to study a foreign language [Galvy 2016].

Because of fast obsolescence, learning the specifics of a particular programming language or computing platform today will not be good job training for tomorrow.

Anyway, there are nowhere near enough teachers for implementing coding for all (and there won't be enough as long as teachers' salaries remain far lower than those of coders and others in IT). Where could the needed money come from? (Hint: Not from a school's sports programs.)

Computer science is a liberal art [Jobs 1995] because to write a computer program involves managing complexity, much as the author of a book or the manager of a project must. You don't write a million-line computer program by writing the first line, then the second, and so on; and you don't do it all by yourself. What we emphasize in teaching computer science is not learning the ins and outs of the syntax and semantics of a particular programming language, but rather cultivating the art of solving a problem by breaking it down into manageable chunks that work together. But that's not simple coding.

Students find learning computer programming to be demanding (logic and syntax must be correct); frustrating (the logic can be wrong, and the computer demands perfect syntax); time-consuming (unlike a term paper that is written the night before it is due, a program is rarely "done" on the first try); but potentially fun (thanks to toy robots and easy-to-program graphics and animation).

Computer science is the science of information transfer. Its key question is, what can information-transfer machines do? Programming (coding) is making machines do those tasks. So, what students really need to learn is how to get machines to do what they want—even as the machines change with advances in technology.

What makes for a good programmer—and what should go into coding for all—may indeed be educationally valuable for all: logical thinking; persistence (since the first version of a program is unlikely to work); algorithmic thinking (ordering steps); following instructions carefully (so a program does what it is supposed to do); contingency planning ("since the user may type in rubbish, we need to validate the input"); thoroughness (the program should be correct and "bullet-proof"); attention to detail; and justifying correctness by sticking to standards (e.g., documentation of code).

But are these qualities better taught through other means (e.g., art, music, mathematics, or sports)? Is there any transfer of learning from one domain to another? Are some qualities mainly elements of "character"? If so, can they be taught, cultivated, or enhanced?

No Evidence for Elementary Mathematics

We would expect that with more than 2,000 years of experience, by now the benefits of learning mathematics would be established in concrete terms, well-known and recognized by all. Unfortunately, that is not the case; there is little research about those benefits, how they are realized, or how best to achieve them, in terms of pedagogy or curriculum. Why so?

The lack may be in part a consequence of concentration on the demonstrable and well-recognized applied benefits of "math skills," such as arithmetic for commerce. It may also be because only in recent times has pragmatism become a major educational philosophy, accompanied by a business model for mass public education. And unlike "math skills," the habits of mind putatively fostered by the study of mathematics may be hard to formulate, agree on, and measure.

Although language study or mathematics may develop desired habits of mind applied in that subject matter specifically, modern psychological studies provide convincing evidence against general transfer to other domains: Transfer is unlikely beyond the domain of initial learning, in part because knowledge is largely domain-specific [Detterman 1993].

Johnson [2012] finds no evidence that studying mathematics enhances logical thinking:

> There appears to be no research whatsoever that would indicate that the kind of reasoning skills a student is expected to gain from learning algebra would transfer to other domains of thinking or to problem solving or critical thinking in general.

Inglis and Attridge [2017] too fail to support the Theory of Formal Discipline, though they suggest that studying "advanced" mathematics is "associated with development of reasoning skills," particularly "the ability to reject invalid inferences."

Another perspective is that perhaps the explanation for the association between mathematics and habits of mind goes in the opposite direction:

> A major reason why it looks as if good thinkers have been helped by taking certain school studies is that there is an inherent tendency of the good thinkers to take such courses. When the good thinkers studied Greek and Latin, these studies seemed to make good thinking.
>
> [Clark 2011]

This *filtering hypothesis*, favored by psychologists, suggests that better reasoners do better in mathematics (or Latin, or chess, or programming), do more of it, and become even better at it. There may be a lack of evidence for elementary mathematics—so far. Nevertheless, learning mathematics may in fact develop good thinking.

Let us consider stages in contemporary mathematics education and their potential for learning how to think.

ARITHMETIC, ALGEBRA, GEOMETRY, TRIGONOMETRY

Arithmetic involves memorizing facts (such as times tables), memorizing algorithms for calculation, inculcating calculational rules (e.g., commutativity of addition and multiplication), identifying patterns, and applying techniques for approximation.

Algebra offers much in the same vein and in particular expands on application and manipulation of symbols.

Geometry reasons about idealized entities in space. There are lots of terminology (e.g., congruence of triangles) and many facts to learn and apply (e.g., side-angle-side for congruence). Analytic geometry connects to algebra by translating back and forth from geometric facts to algebraic ones. Synthetic geometry, the logical deduction of theorems from axioms, used to be the heart of the study of geometry, but many contemporary high school courses in geometry ask students to do little if any proving on the grounds that it is "too hard for the average student."

Trigonometry, college algebra, and precalculus largely practice and develop further proficiency in algebra and other skills.

It may be that "mathematical training strengthens the mind . . . is as impossible to prove as the proposition that music and art broaden and enrich the soul" [Dudley 1997, 363–364]. But perhaps more-evident training of mind takes place in further mathematics courses?

What about Calculus?

For a century or more, the mathematics curriculum has funneled toward calculus, sieving out students along the way. As taught . . .

. . . **calculus is computation.** The theory of calculus—its theorems, lemmas, proofs, and algorithms—has a marvelous logical structure, akin to Euclid's for geometry. But students don't get to see the forest for the trees: It is a rare textbook that gives a diagram explaining the dependence of theorems on one another and on the concepts (definitions) and axioms.

. . . **calculus is crass.** Calculus is inextricably involved with fundamental questions about the nature of number and space. But those ideas have been filtered out. Of course, after more years of advanced mathematics, a small remnant of the students gets to see those ideas.

. . . **calculus is condensed.** Over the centuries, calculus has weathered many crises, philosophical and mathematical—George Berkeley's objections [1734] to derivatives involving "ghosts of departed quantities," the conundrums resolved by l'Hôpital's rule for indeterminate limits, and later the nature of a function. But those questions, which helped shape calculus and still puzzle students today, are left out. Only their answers—the "ghosts of departed questions"—remain.

. . . **calculus is not relevant to student interests and aspirations.** My wife formerly taught English as a second language to ninth-grade Spanish-speaking students. To enhance their educational aspirations, she showed the film *Stand and Deliver* [Menéndez and Musca 1988], based on a true story of teaching calculus to dropout-prone Hispanic-heritage students. She asked me for a calculus book that would show the students what the subject is about and what it is useful for. I sadly realized that most calculus books would be neither exciting nor inspirational. I settled on Hahn [1998] (partly because it weighs less than five pounds, and despite the fact that it is not in color). It starts from Greek astronomy, considers ballistics, has a section entitled "Incredible Consequences" about Newton's investigations, and discusses suspension bridges, telescopes, nuclear decay, the Earth's population, rocket propulsion, and the price of oil. (Hahn [2017] is a newer work in the same spirit.)

In the 1990s, there was a calculus reform that reconsidered the pedagogy of calculus by adopting new instructional techniques, applications content, and technology use.

My concern is not instructional techniques. Instead, I offer an expansion on the points above to offer a philosophical critique, with how to teach calculus situated in the mainstream of intellectual pursuits.

Audiences for Calculus

For more than a dozen years, calculus has been mainly a high school subject: More students take it—in some version—in high school than in college [Bressoud 2004].

Many students take calculus in high school as "college application" mathematics, hoping to enhance chances of admission to preferred colleges. In addition, taking it as an Advanced Placement course can shorten time in college and the resulting expense. Those motivations do not generally result in such students taking further mathematics in college.

Students with strong interest in mathematics or physical science tend to take calculus in high school; hence, few of them appear in introductory calculus in college.

So who takes calculus in college? By and large, students whose newly realized major and career aspirations require calculus. The students did not have the opportunity, the interest, or a strong enough background in mathematics to take calculus in high school. They are unlikely to be interested in mathematics for its own sake, and they are hard to charm with the "romance" of mathematics. Their motivation is further eroded by the fact that apart from physics, economics, engineering, and mathematics itself, disciplines that require calculus of majors rarely require it as a prerequisite for specific courses.

What Is Missed

Notably missing in students' written work in calculus are *words*. There is no explicit link expressed between what they are thinking and the mathematical symbols that they write—they rarely write words! They have not been taught to write down descriptions of the steps from one mathematical expression to the next (e.g., "Now we divide both sides of the equation by *x*, taking care to be sure that *x* cannot be zero"). Mathematics instruction could benefit from following the example of computer science, which demands "commenting" of code; but calculus is no doubt too late a starting point for that needed revolution.

In a similar spirit of disjunction between prose and mathematical expression, and despite years of addressing "word problems," students have almost no ability to translate a verbal description of a situation into a mathematical form.

In other words, despite 12 years of "mathematics" instruction, they have not developed an ability to translate among domains (words, symbols, figures)—a key ingredient of mathematical thinking.

Calculus Is Computation

Calculus is about modeling continuous change, chiefly by means of linear approximations.

But we teach calculus as techniques, to calculate limits, derivatives, and antiderivatives of elementary functions by applying a small number of rules (algorithms) to functions built recursively (a word we dare not mention) from simpler functions. Students find areas, volumes, and optima for situations described by such functions. They determine whether series converge (but not often to what) and expand functions into power series.

Is proficiency in calculus computation really necessary? Absolutely, for prospective engineers, physical scientists, economists, mathematicians, and perhaps computer scientists. Other students would benefit more from an intellectually enriched calculus, or a different experience featuring more-mathematical thinking, or even some other computational course.

Despite the emphasis in calculus on calculational algorithms (which we and students call *formulas*), we don't discuss algorithmic aspects, much less efficiency. The class of elementary functions is closed under differentiation; but integration applied to them can lead out of that class, to the normal error function and elliptic functions, which are no less important because they cannot be expressed in elementary terms. There is indeed an algorithm to decide whether a given elementary function has an elementary indefinite integral and to produce it if it exists [Risch 1970]. The algorithm is intricate, so basically we throw heuristics for integration at students and let them gain experience by working through most of the kinds of problems that may come up.

Thus, we miss an opportunity to add to students' perception of mathematics: Not only can mathematics establish easily believed existence results (such as the Extreme Value Theorem, the Intermediate Value Theorem, and the Mean Value Theorem), but it can also prove impossibility results (e.g., the integral $\int e^{x^2}\, dx$ cannot be expressed as an elementary function).

"It's not 'cookbook'—we give proofs." Mathematics instructors pride themselves that their teaching is more than "plug and chug" (plugging numbers into formulas and working out the results), a practice sometimes derogatorily attributed to the local engineering school. We work through proofs in front of students to justify that the formulas and rules are right. In some cases, it's essential that we convince students why and how the world is more complicated than they would like to believe—for example, why the derivative of a product is not the product of the derivatives.

Although mathematicians rightly take pride in supporting skills with proofs, rarely do we assign "problems to prove" rather than "problems to find" [Pólya 1945]. Why? Because after 12 years of mathematics-as-computation, most students are not up to offering a mathematical argument.

Some of the computation in calculus is suitable to an earlier era. Here are just four examples of calculus topics that should go, since they distract from the dominant idea of calculus (linearization of change), and even from its perverted purpose of computation, and contribute neither to its intellectual content nor to mathematical thinking.

Calculus for Curve Sketching: In an era of graphing calculators, calculating derivatives to help sketch a curve is simply superfluous.

Tests for Extrema: The great achievement of calculus in optimization is to reduce the candidates for maxima and minima from all real numbers to a small finite number—in other words, to convert the problem to a discrete problem. At that point, just *try* all of the (very few) candidates. Or just input the function to a graphing calculator in the first place.

Simpson's Rule: Calculus is fundamentally about linear approximation of functions. There's no need to bring in second-order approximations for their small gain in efficiency—a calculator can quickly get as accurate a result as needed from summing rectangles.

Tests for Convergence of Series: The reason for infinite series in calculus is to approximate functions by polynomials (Taylor's theorem). The ratio test suffices to determine where the series works. All the numerical series (harmonic series, p-series, etc.) and the tests (root test, limit comparison test, etc.) are beautiful mathematics. But it's mathematics that has nothing at all to do with the central focus of calculus, the modeling of change.

Calculus Is Crass

What's a limit? The key concepts of calculus, the derivative and the integral, are (different) kinds of limits; but textbooks do not always offer a definition to refine the intuition of what a limit is.

What's an infinitesimal? Are there any? Does it matter? Calculus books in a former era were titled *Infinitesimal Calculus*, to distinguish the subject from logical calculus.

Times have changed. In most calculus textbooks today, the word *infinitesimal* does not occur! What we have lost, physicists still retain, though their courses too avoid key questions: Do space and time occur in infinitesimal quantities? Do infinitesimal numbers exist? Even if they don't, are infinitesimal numbers useful—does it make physical and mathematical sense to calculate as if they do? ("Of course," say the physicists.) If so, how?

The question of infinitesimals was revolutionary in mathematics, and our students should have an opportunity to realize it. Were we mathematicians so scored by Berkeley's critique of limits ("ghosts of departed quantities" [1734]), and so unable to answer the criticism for 200-plus years, that we are afraid to broach the subject? Or did Abraham Robinson, who proved in the 1960s that it's logically consistent to extend the real numbers to include infinitesimals and reason with them [1966], make a fatal error in public relations by calling such reasoning *nonstandard analysis*? Or do we just regard these questions as irrelevant to the mission of calculus as computation?

Calculus Is Condensed

Where is the blindered horse calculus going, as it plods through calculations? And what countryside escapes its purview?

That countryside is easy to identify but more difficult to teach about than calculation: human struggles with ideas, history, social and cultural context; ontogeny of ideas; a sense of mathematicians as people—not to mention applications as inspiration and driving force.

To change metaphors, calculus is condensed to a black hole of computation—reduced content, easy to teach, easy to evaluate learning, hard to defend as an intellectual activity.

The results of calculus depend on the real numbers having properties that students are told are reasonable but which at some point deserve fuller investigation. The overwhelming majority of students will never take another mathematics course, so calculus is their last chance to learn about numbers. What indeed is a number? Can we imagine different kinds [Knuth 1974, Henle 2013]?

We note several crises in the foundations of calculus: the discovery of irrational numbers (400 BC?); the question of the existence of infinitesimals (1734); confusion about convergence of infinite series (1807); the unknown size of the collection of real numbers (the continuum)—how many real numbers are there?—investigated by Georg Cantor in the 1880s; the challenge of the discrete by computer science in the 1980s. Shouldn't students learn about some of these great developments before they finish their last mathematics course ever?

Calculus Is Irrelevant

Textbook applications of the derivative are either related rates or max/min problems. The former concern containers in familiar geometric shapes filling or emptying, vehicles passing the same point at different times, shadows growing or shrinking, change in angle as an object moves nearer or farther away, sliding ladders, and or melting snowballs. Max/min problems treat objects dropped or thrown upward (but only in a vacuum), fencing a field, making a box or can, stiffness of a beam, carrying a pole around a corner, profit (usually with fictitious unrealistic price and cost functions), time or cost of path to a goal through differing media (e.g., Snell's law), and various geometric optimizations (e.g., largest cylinder inscribed in a sphere).

Textbook applications of the integral are scarcer, in part because they may lead to integrals that must be evaluated numerically (and hence do not practice students in integration techniques): rectilinear motion; lengths, areas, and volumes of geometric objects; and work, fluid force, center of mass, and moments.

Of course, it is delightful that calculus can solve all these problems. But are these really "practical problems" for our students? Well, such problems indeed reinforce mathematics learned earlier: graphing, the Pythagorean theorem, similar triangles, trigonometry. They can be "practical" in giving students *practice* in interpreting a problem into mathematical notation, and in differentiating and integrating various functions clothed in a semblance of connection to reality. It's just not most students' reality.

Dudley [1988] examined 85 calculus books, including all of their applications of maxima and minima. He concluded that calculus is too hard to teach to college freshmen, and first-semester calculus has *no* applications ("artificial problems provide valuable practice in translating from English into mathematics and that is all they are for").

Should calculus have to be relevant? to what? to whose experience? On another educational front, is literature relevant? history? philosophy? Should they have to be?

Nothing of the critique above dishonors calculus as a magnificent human achievement, worth studying and knowing for its own sake as well as for interpreting the world around us.

Calculus should emphasize and epitomize the structure of mathematical thinking: distill intuition into precise definitions, reason to general facts (theorems), and apply theorems to real situations.

But structured thinking is not why students are required by other departments to take calculus, and not what they desire to get out of it. Students are victims of the pragmatism of American education, with its orientation toward commerce and accompanying encouragement of personal economic anxiety keyed to educational attainment level.

Nevertheless, students emerging from calculus should be able to explain to friends and relatives what calculus is about, as an endeavor of human thought, and also what it can be good for, in practical terms. We need to do better in helping students give satisfying answers.

Let's "Intellectualize" Calculus . . .

How do you measure intellectual growth and development? We can measure knowing facts but not acquiring values, developing interests, or enhancing sensitivities. Education is not training, as the arts and humanities have long realized.

What should we do to make training in calculus also *educational*?

Teach calculus in a way that is philosophically sound and intellectually defensible. Intellectual honesty does not demand that we prove everything but requires us to point out what we sweep under the rug—e.g., the axioms about real numbers that we assume and what we are accepting without offering proof (such as a well-founded definition of a limit or of area).

Embed calculus in historical, cultural, scientific, and applications contexts. Why was it developed when it was, in ancient Greece, in seventeenth-century Europe, and independently in Japan? (See Otero [1999] and Bressoud [2019].)

Place calculus in the context of history of computational capability and adapt it to current capabilities.

Come to grips with infinitesimals.

. . . and *"Pragmatize"* Calculus

Calculus is a humanity, but it is also an applied subject.

Center the course around modeling change, starting from differential equations and difference equations.

Stick to the main story—the analysis of change—despite favored topics that have accreted over the years.

Serve the majority, not mainly potential majors in pure mathematics. Make the course complete in itself but beckoning toward more, and postpone computational skill building to a second course for those who need such skill.

Reduce emphasis on analytic solutions in favor of computer-aided numerical solutions.

Get rid of unreal examples and exercises; they are demotivating. Sample from a calculus book: "Two fleas are arguing about who will get the longest ride when Jenny pedals her tricycle home from the park . . . ."

Intellectualizing calculus demands that faculty (and particularly teaching assistants) know better the history of calculus. Pragmatizing calculus forces them to connect the subject to both ancient and modern uses—that is, to know more science.

Such a two-pronged transition would indeed let learning calculus become a part of training the mind. A further reward would be to integrate not just calculus but also mathematicians themselves—outcasts for too long!—into the intellectual community.

Acknowledgment

This essay is adapted from Campbell [2006, 2016, 2017, 2018].

References

Aleven, Vincent. 2012. Learning and transfer. http://learnlab.org/opportunities/summer/presentations/2012/TransferSumSch2012.pdf.

Berkeley, George. 1734. *The Analyst; or, a Discourse Addressed to an Infidel Mathematician*. London: J. Tonson. 2004. Reprinted. Whitefish, MT: Kessinger Publishing; also available as an e-book. Together with two subsequent essays in response to mathematicians, in Alexander Campbell Fraser, *The Works of George Berkeley*. Oxford, U.K.: Clarendon Press, 1901.

Bressoud, David M. 2004. The changing face of calculus: First-semester calculus as a high school course. *Focus* (August–September) 6–8. 2010. Reprinted in Diefenderfer and Nelsen [2010], 54–58. https://www.maa.org/the-changing-face-of-calculus-first-semester-calculus-as-a-high-school-course.

———. 2019. *Calculus Reordered: A History of the Big Ideas*. Princeton, NJ: Princeton University Press.

Burger, Edward B. 2019. *Making Up Your Mind: Thinking Effectively through Creative Puzzle-Solving*. Princeton, NJ: Princeton University Press.

Campbell, Paul J. 2006. Calculus is crap. *The UMAP Journal of Undergraduate Mathematics and Its Applications* 27 (4) (2006): 415–430.

———. 2016. STEM the tide? *The UMAP Journal of Undergraduate Mathematics and Its Applications* 37 (1) (2016) 1–7.

———. 2017. Calculus is irrelevant. *The UMAP Journal of Undergraduate Mathematics and Its Applications* 38 (4) (2017): 355–368.

———. 2018. Mathematics: Unreasonably ineffective? *The UMAP Journal of Undergraduate Mathematics and Its Applications* 39 (1) (2018): 1–4.

Clark, Donald R. 2011. Transfer of learning. http://www.nwlink.com/~donclark/hrd/learning/transfer.html.

Code.org. 2016. Where computer science counts. https://code.org/action.

Detterman, D. K. 1993. The case for the prosecution: Transfer as an epiphenomenon. In *Transfer on Trial: Intelligence, Cognition, and Instruction*, edited by D. K. Detterman and R. J. Sternberg, 1–24. Westport, CT: Ablex.

Diefenderfer, Caren L., and Roger B. Nelsen (eds.). 2010. *The Calculus Collection: A Resource for AP and Beyond*. Washington, DC: Mathematical Association of America.

Dudley, Underwood. 1988. Book review: *Calculus with Analytic Geometry*. By George Simmons. *American Mathematical Monthly* 95 (9) (November 1988): 888–892. Reprinted in Diefenderfer and Nelsen [2010], 69–73.

———. 1997. Is mathematics necessary? *College Mathematics Journal* 28 (5) (November 1997): 360–364.

———. 2008. Calculus isn't crap. *The UMAP Journal* 29 (1): 1–4.

Galvy, Gaby. 2016. Some say computer coding is a foreign language. http://www.usnews.com/news/stem-solutions/articles/2016-10-13/spanish-french-python-some-say-computer-coding-is-a-foreign-language.

Hahn, Alexander J. 1998. *Basic Calculus: From Archimedes to Newton to Its Role in Science*. New York: Springer.

———. 2017. *Calculus in Context: Background, Basics, and Applications*. Baltimore: Johns Hopkins University Press.

Henle, Michael. 2013. *What Numbers Are Real?* Washington, DC: Mathematical Association of America.

Inglis, Matthew, and Nina Attridge. 2017. *Does Mathematical Study Develop Logical Thinking? Testing the Theory of Formal Discipline*. Hackensack, NJ: World Scientific.

Jobs, Steve. 1995. Steve Jobs on computer science. https://www.youtube.com/watch?v=IY7EsTnUSxY.

Johnson, Peter. 2012. Does algebraic reasoning enhance reasoning in general? A response to Dudley. *Notices of the American Mathematical Society* 59 (9) (October 2012): 1270–1271. http://www.ams.org/notices/201209/rtx120901270p.pdf.

Knuth, Donald E. 1974. *Surreal Numbers: How Two Ex-Students Turned on to Pure Mathematics and Found Total Happiness*. Reading, MA: Addison-Wesley.

Menéndez, Ramón (director), and Tom Musca (producer). 1988. *Stand and Deliver*. Color film, 99 min. Script by Ramón Menéndez and Tom Musca. Starring Edward James Olmos and Estelle Harris. Hollywood, CA: American Playhouse, Warner Bros. Pictures.

Otero, Daniel E. 1999. Calculus from an historical perspective: A course for humanities students. *PRIMUS (Problems, Resources, and Issues in Undergraduate Mathematics Studies)* 9 (1): 56–72.

Pólya, George. 1945. *How to Solve It: A New Aspect of Mathematical Method*. 1948. 2nd ed. 2004. Reprint. Princeton, NJ: Princeton University Press.

Risch, R. 1970. The solution of the problem of integration in finite terms. *Bulletin of the American Mathematical Society* 76: 605–608.

Robinson, Abraham. 1966. *Nonstandard Analysis*. New York: North-Holland. 1974. Rev. ed. New York: American Elsevier.

Abstracting the Rubik's Cube

David Hilbert wrote, "The art of doing mathematics consists in finding that special case that contains all the germs of generality."

Over the past few decades, a growing group of puzzle enthusiasts known as *hypercubists* have generalized the Rubik's Cube in ways that traverse a wide expanse of mathematical ground. The explorations have been a microcosm of mathematical progress. Finding and studying these puzzles provides a rich way to approach varied topics in mathematics: geometry (higher dimensional, non-Euclidean, projective), group theory, combinatorics, algorithms, topology, polytopes, tilings, honeycombs, and more.

For this group of people, twisty puzzles are more than just a casual pastime. Elegance is a core principle in their quest.

FIGURE 1. The Rubik's cube, warped! See also color images.

Hypercubes

We can change many properties of the classic $3 \times 3 \times 3$ Rubik's Cube, such as its shape or twist centers, to make new and interesting puzzles (Figure 2). But the hypercubing group began by changing a more abstract property, namely, the dimension.

Don Hatch and Melinda Green created an exquisite working four-dimensional $3 \times 3 \times 3 \times 3$ (or 3^4) analogue, which they called Magic-Cube4D. Every property of this puzzle is upped a dimension: Faces, stickers, and twists are three-dimensional rather than two-dimensional. Figure 3 shows the ordinary Rubik's Cube and the hyperpuzzle using a central projection that reduces the dimension by one; it is as if we are looking into a box, with the nearest face hidden.

The 3^3 Rubik's Cube has $6 \times 3^2 = 54$ stickers that can live in a mind-boggling 4.325×10^{19} possible states. The hypercubical 3^4 has $8 \times 3^3 = 216$ stickers, and the number of possible puzzle positions explodes to an incomprehensible 1.756×10^{120}. Calculating this number is a challenge that will test your group theory mettle!

But as Edwin Abbott wrote in *Flatland*, "In that blessed region of Four Dimensions, shall we linger on the threshold of the Fifth, and not enter therein?" The group didn't stop at four dimensions. In 2006, a working five-dimensional puzzle materialized with $10 \times 3^4 = 810$ hypercubical stickers and 7.017×10^{560} states, pushing the boundaries

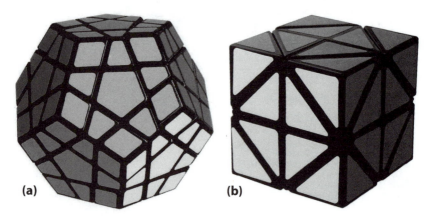

(a)   **(b)**

FIGURE 2. (a) Megaminx uses a dodecahedral shape rather than a cube. (b) The Helicopter Cube twists around edges instead of faces. See also color images.

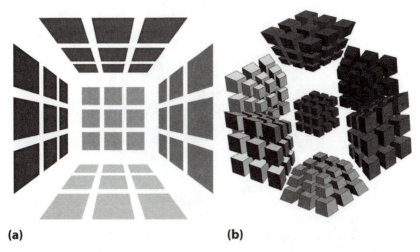

(a) **(b)**

FIGURE 3. Projection tricks can help us visualize higher dimensional Rubik's Cubes. (a) The 3^3, projected so that a two-dimensional "flatlander" sees five of the six cube faces. (b) The 3^4, projected so that a three-dimensional being sees seven of the eight hypercube faces. See also color images.

FIGURE 4. A shadow of a shadow of a shadow of the 3^5. Stickers are little hypercubes. See also color images.

of visualization. Figure 4 shows a shadow of a shadow of a shadow of the five-dimensional object. Nonetheless, as of mid-2017, around 70 people have solved this puzzle.

In June 2010, Andrey Astrelin stunned the group by using a creative visual approach to represent a seven-dimensional Rubik's Cube. Yes, it has been solved. Can you calculate the number of stickers on the 3^7?

FIGURE 5. The Magic120Cell, or the 4D Megaminx, has 120 dodecahedral faces. It derives from the 120-cell, one of six Platonic shapes in four dimensions. See also color images.

You may also enjoy trying to work out the properties of a two-dimensional Rubik's Cube. What dimension are the stickers?

Of course, we can play the shape-changing game in higher dimensions too, yielding a panoply of additional puzzles. There are five Platonic solids in three dimensions, but six perfectly regular shapes a dimension up, and you can attempt to solve twisty puzzle versions of all of them! Figure 5 shows one of the most beautiful in its pristine state.

Shapes in arbitrary dimensions are called *polytopes*, or *polychora* in four dimensions. In addition to the regular polychora, there are many uniform polychora, and quite a few have been turned into twisty puzzles. Uniform polychora can break regularity in various ways. They may have multiple kinds of three-dimensional faces, or the faces may be composed of uniform (that is, Archimedean) polyhedra.

Curved Twisty Puzzles

"For God's sake, I beseech you, give it up. Fear it no less than sensual passions because it too may take all your time and deprive you of your health, peace of mind and happiness in life."

No, these were not desperate pleas to a hypercubist about excessive puzzling adventures. Such were the words of Farkas Bolyai to his

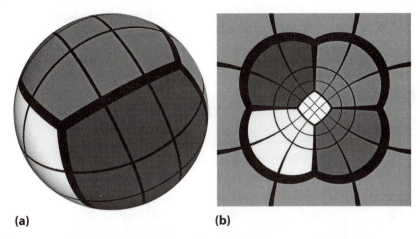

(a) **(b)**

FIGURE 6. (a) The Rubik's Cube projected radially onto a sphere yields a two-dimensional tiling of the sphere. (b) It is then stereographically projected onto the plane. See also color images.

son János, discouraging him from investigating Euclid's fifth postulate. János continued nonetheless, which led him into the wonderful world of hyperbolic geometry.

We will also not heed the elder Bolyai's advice. Let's use topology to abstract away a different property of Rubik's Cube—its cubeness. To do so, project the cube radially outward onto a sphere (Figure 6a). Notice that all the important combinatorial properties remain. Furthermore, what were planar slices of the Rubik's Cube are now circles on the sphere's surface. A twist simply rotates the portion of the surface inside one of these twisting circles. In short, we are viewing the Rubik's Cube as a tiling of the sphere by squares, sliced up by circles on the surface.

Inspired by this example, we can consider other colored regular tilings, and a huge number of new twisty puzzles become possible, some living in the world of hyperbolic geometry!

For two-dimensional surfaces, there are three geometries with constant curvature: spherical, Euclidean, and hyperbolic. These geometries correspond to whether the interior angles of a triangle sum to greater than, equal to, or less than 180 degrees, respectively. Intuitively, we can think of the surface of a sphere, a flat plane, and a Pringles potato chip as representative surfaces for these geometries.

Each surface of constant curvature can be tiled with regular polygons. The *Schläfli symbol* encodes regular tilings with just two numbers, $\{p,q\}$. This denotes a tiling by p-gons in which q such polygons meet at each vertex. The value $(p-2)(q-2)$ determines the geometry: Euclidean when equal to 4, spherical when less, and hyperbolic when greater.

For example, $\{4,3\}$ denotes a tiling by squares with three arranged around each vertex, that is, the cube. As we saw in Figure 6a, this arrangement gives a tiling of the sphere, and indeed, $(4-2)(3-2) = 2 < 4$.

Euclidean geometry is the only one of the three geometries that can live on the plane without any distortion. A lovely way to represent the other geometries on the plane is via *conformal*, or angle-preserving, maps. The *stereographic projection* is a conformal map for spherical geometry. Figures 1 and 6b show the stereographic projection of the spherical Rubik's Cube onto the plane. For hyperbolic geometry, we use the *Poincaré disk*, which squashes the infinite expanse of the hyperbolic plane into a unit disk (Figure 10).

One challenge of turning Euclidean and hyperbolic tilings into twisty puzzles is that unlike spherical tilings, which are finite, tilings of these two geometries go on forever. To overcome this hurdle, we begin with a tiled surface, called the *universal cover*; choose a certain subset of tiles, called the *fundamental domain*; and *identify* its edges to form a *quotient surface*. Intuitively, we glue the edges of this region together to turn the infinite tilings into finite puzzles. Figures 7, 8, and 9 show a few examples.

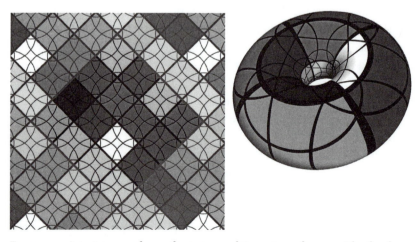

FIGURE 7. A twisty puzzle on the torus and its universal cover. The fundamental domain is outlined in red. See also color images.

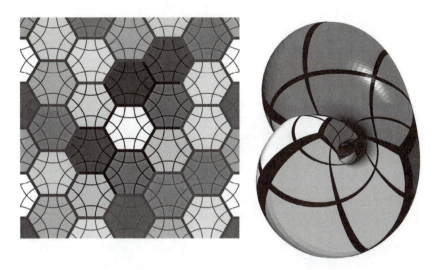

FIGURE 8. A twisty puzzle on the Klein bottle and its universal cover. The fundamental domain is outlined in red. See also color images.

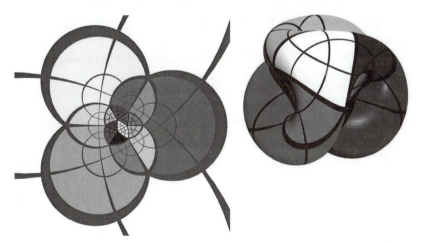

FIGURE 9. A twisty puzzle on Boy's surface (the real projective plane) and its universal cover. The fundamental domain is outlined in red. See also color images.

One of the crown jewels of this abstraction is the *Klein quartic* Rubik's Cube, composed of 24 heptagons, three meeting at each vertex. It has "center," "edge," and "corner" pieces, just like the Rubik's Cube. The universal cover is the $\{7,3\}$ hyperbolic tiling, and the quotient surface is a three-holed torus. This puzzle contains some surprises; if you

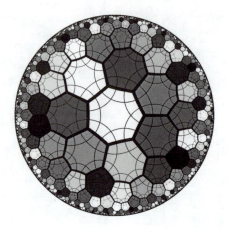

FIGURE 10. Klein quartic Rubik's Cube on the hyperbolic universal cover. The quotient surface is a three-holed torus. See also color images.

solve it layer by layer, as is common on the Rubik's Cube, you'll be left with two unsolved faces at the end instead of one.

All these puzzles and more are implemented in a program called MagicTile. The puzzle count recently exceeded a thousand, with an infinite number of possibilities remaining.

More Puzzles

There are even more intriguing analogues that we have not yet seen. Let me mention two of my favorites. The first is another astonishing set of puzzles by Andrey Astrelin based on the {6,3,3} *honeycomb* in three-dimensional hyperbolic space, $\mathbb{H}^3$ (Figure 11). The faces are hexagonal {6,3} tilings, with three faces meeting at each edge. Gluing via identifications serves to make the underlying honeycomb finite in two senses: the number of faces and the number of facets per face. If we take a step back and consider where we started, this puzzle has altered the dimension, the geometry, and the shape compared to the original Rubik's Cube!

The second is a puzzle created by Nan Ma based on the 11-cell, an abstract regular polytope composed of 11 hemi-icosahedral cells (Figure 12). This is a higher dimensional cousin of the Boy's surface puzzle in Figure 9. The 11-cell can only live geometrically unwarped

FIGURE 11. An in-space view of the Magic Hyperbolic Tile {6,3,3} puzzle in three-dimensional hyperbolic space. See also color images.

FIGURE 12. The scrambled Magic 11-Cell. See also color images.

in 10 dimensions, but Nan was able to preserve the combinatorics in his depiction.

So many puzzles have been uncovered that one could be forgiven for suspecting there is not much more to do. On the contrary, there are arguably more avenues to approach new puzzles now than 10 years ago. For example, there are no working puzzles in $\mathbb{H}^3$ composed of finite polyhedra. There are not yet puzzles for uniform tilings of Euclidean or hyperbolic geometry, in two or three dimensions. Uniform tilings are not even completely classified, so further mathematics is required before some puzzles can be realized. Melinda Green has been developing a physical puzzle that is combinatorially equivalent to the 2^4. The idea of fractal puzzles has come up, but no one has yet been able to find a good analogue.

In addition to the search for puzzles, countless mathematical questions have been asked or are ripe for investigation. How many permutations do the various puzzles have? What checkerboard patterns are possible? Which n^d puzzles have the same number of stickers as pieces? How many ways can you color the faces of the 120-cell puzzle? What is *God's number* for these higher dimensional Rubik's Cubes; that is, what is the minimum number of moves in which the puzzle can be solved, regardless of starting position? The avenues are limited only by our curiosity.

As John Archibald Wheeler wrote, "We live on an island surrounded by a sea of ignorance. As our island of knowledge grows, so does the shore of our ignorance."

Further Reading

The *MagicCube4D* website (superliminal.com/cube/cube.htm) contains links to all the puzzles in this article and to the hypercubing mailing list.

Burkard Polster (Mathologer) produced wonderful introductory videos to MagicCube4D and MagicTile "Cracking the 4D Rubik's Cube with simple 3D tricks" (youtu.be /yhPH1369OWc) and "Can you solve THE Klein Bottle Rubik's Cube?" (youtu.be /DvZnh7-nslo).

The following papers are freely available online:

H. J. Kamack and T. R. Keane, "The Rubik Tesseract," (1982) http://bit.ly/RubikTess.

John Stillwell, "The Story of the 120-cell," *Notices of the AMS* 48, 1 (2001): 17–24.

Carlo H. Séquin, Jaron Lanier, and UC CET, "Hyperseeing the Regular Hendecachoron," *Proc. ISAMA* (2007): 159–166.

Topology-Disturbing Objects:
A New Class of 3D Optical Illusion

Kokichi Sugihara

1. Introduction

The topology of a geometric object provides the most fundamental properties, in the sense that they are not disturbed under any continuous transformation; for example, the connected components remain connected, and disconnected ones remain disconnected. Topological properties are thus preserved when objects are rotated around a vertical axis because the rotation is one of the simplest continuous transformations. However, we found a class of objects for which the topological properties appear to be disturbed when they are seen from two viewpoints. Of course, physically, this cannot happen, but an optical illusion can make humans perceive that it has happened.

An effective method for presenting two views simultaneously is to use a mirror. By placing an object and a mirror appropriately, we can see the object and its reflection in the mirror, and thus we can compare the two appearances. If the topologies are different from each other, we might have a strong sense of wonder and impossibility. Thus, we can say that the topology-disturbing objects are those whose topologies appear to change in a mirror.

In the history of arts and entertainment, optical illusions have been used to create the impression of impossibility. Penrose and Penrose [10] and Escher [1, 5] drew pictures of impossible objects, which are optical illusions because they appear to represent three-dimensional (3D) structures, but ones that could not possibly exist. Moreover, these impossible objects have been realized as 3D structures by using optical illusions generated by tricks involving discontinuity and curved surfaces [4]. Physically impossible motion can also be created by antigravity slopes, where the orientations of the slopes are perceived to be opposite to their true orientations; in these illusions, balls appear to defy gravity and roll uphill [14, 15]. Another example is Hughes' 3D painting method known as reverse perspective [19]. He painted pictures on a 3D surface in such a way that near objects are painted on a part of the surface that is farther from the viewer, and far objects are painted nearer to the viewer. The result is that we perceive unexpected motion when we move our heads [3, 9]. A similar depth-reversal trick, known as the *hollow-face illusion*, is used in haunted mansions to create a visual impression that the gaze of a statue is following the viewer [6, 13]. Recently, Sugihara found ambiguous cylinders that appear drastically different when viewed from two specific viewpoints [16, 17]. The objects presented in this paper are variants of ambiguous cylinders. Specifically, the original ambiguous cylinders disturb the geometry, and the present objects disturb the topology.

The visual effect of topology change can be considered a variant of traditional anamorphosis. In *anamorphosis*, painting on a plane or on a 3D surface looks meaningless when seen from a general viewpoint, but it becomes meaningful when seen from a single specific viewpoint.

A typical example is Gregory's 3D realization of the Penrose impossible triangle. The Penrose impossible triangle is an anomalous picture of an imaginary 3D structure that is evoked in our brain when we see the picture but that cannot be constructed physically [10]. Gregory [6] created a 3D model of an open path composed of three rods that appear to close into the Penrose impossible triangle when seen from a special viewpoint. This is an example of anamorphosis because it looks like nothing in particular when seen from a general viewpoint but looks like an impossible triangle when seen from a unique special viewpoint.

Note that traditional anamorphosis gives meaningful appearance when it is seen from one special viewpoint. The topology-disturbing objects, on the other hand, are accompanied by two specific viewpoints, from which they appear to be meaningful but drastically different, and thus they create the sense of impossibility. From this aspect of their nature, we might consider the topology-disturbing objects as a kind of multiple anamorphosis. Note that there are various other classes of multiple anamorphoses. One class is the multiple-silhouette sculptures, such as *Encore* (1976) by Shigeo Fukuda, which looks like a silhouette of a pianist and a silhouette of a violinist when seen from two special viewpoints, and *1, 2, 3* by James Hopkins, which gives three silhouettes of "1," "2," and "3." Another class is the multiple-appearance wire frame art such as the one that appears to be an elephant and giraffes when seen from two special viewpoints.

We will show examples of topology-disturbing objects (Section 2), briefly review the principle of ambiguous cylinders (Section 3), and apply the principle to the design of topology-disturbing objects (Section 4). Next, we summarize a general condition for the constructability of the topology-disturbing object (Section 5), and give concluding remarks (Section 6). We also present a diagram that shows the unfolded surfaces of a simple example of a topology-disturbing object; from this diagram, one may construct the object by paper crafting (Appendix A). Videos of topology-disturbing objects can also be found on YouTube [18].

2. Examples of Topology-Disturbing Objects

Figure 1(a) shows an example of a topology-disturbing object. The direct view of this object consists of two rectangular cylinders that are separated from each other. A plane mirror is positioned vertically behind the object, and the object can also be seen in the mirror. However, in the mirror, the two cylinders appear to intersect each other. Thus, the direct view and the mirror image have different topologies, which seems impossible.

The mirror is an ordinary plane mirror, and hence it just gives us another view of the object. Therefore, a topology-disturbing object appears to have two different topological structures when viewed from

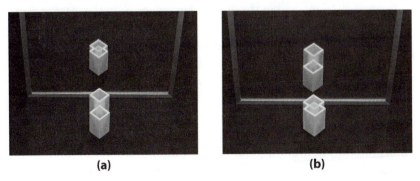

(a)   **(b)**

FIGURE 1. Topology-disturbing pair of rectangular cylinders.

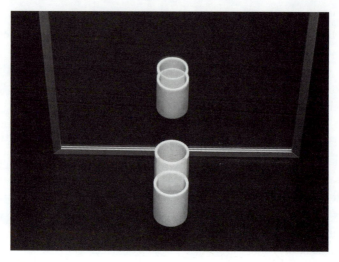

FIGURE 2. Topology-disturbing pair of circular cylinders.

two special directions. Actually, if we rotate the object around a vertical axis by 180 degrees, the direct appearance and the mirror image are interchanged, as shown in Figure 1(b). This behavior is typical of topology-disturbing objects.

If we replace the rectangular cylinders with circular cylinders, we can construct a similar topology-disturbing object, as shown in Figure 2. The direct view consists of two circular cylinders that are separated, but in the mirror view, they intersect.

Figure 3 shows what happens if another cylinder is added, and the three cylinders are placed along a slanted line. As before, the three cylinders are disconnected in the direct view, whereas in the mirror view, they appear to intersect.

Figure 4 shows another type of a topology-disturbing object. In the direct view, the two cylinders are concentric (nested). However, in the mirror, the cylinders appear to have changed shape and to intersect each other.

Figure 5 shows another object, which consists of a cylinder and a plane. In the direct view, the plane passes through the cylinder, whereas in the mirror, the plane and the cylinder are separated.

These examples are typical topology-disturbing objects. Each object consists of two or more cylinders (Figure 5 includes a plane). In one

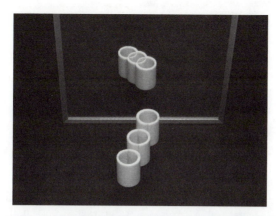

FIGURE 3. Topology-disturbing triplet of cylinders.

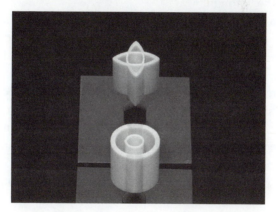

FIGURE 4. Topology-disturbing nesting cylinders.

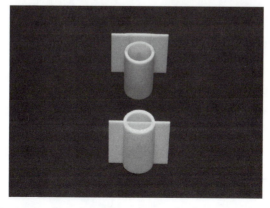

FIGURE 5. Topology-disturbing cylinder and a plane.

view, they are disconnected, whereas in the other view, they intersect. Disconnection and intersection are topologically invariant properties, but in this case, they are not preserved in the mirror image, and thus we call them topology-disturbing objects.

3. Principle of Ambiguous Cylinders

Topology-disturbing objects can be constructed by applying a variation of the method used to design ambiguous cylinders. Hence, in preparation, we briefly review the principle of ambiguous cylinders [16]. An ambiguous cylinder is a cylindrical object that appears to have different structures when viewed from two special directions. They can be constructed in the following way:

As shown in Figure 6, we fix an xyz Cartesian coordinate system so that the xy-plane is horizontal, and the positive z direction orients upward. Let $v_1 = (0, \cos\theta, -\sin\theta)$ and $v_2 = (0, -\cos\theta, -\sin\theta)$ be two viewing directions; they are parallel to the yz-plane, and they are directed downward at the same angle θ, but in opposite directions.

Next, as shown in Figure 7, on the xy-plane, we fix two curves $a(x)$ and $b(x)$ for $x_0 \le x \le x_1$. Note that the initial points $a(x_0)$ and $b(x_0)$ have the same x-coordinate, and similarly, the end points $a(x_1)$ and $b(x_1)$ have the same x-coordinate. Moreover, these two curves are x-monotone. For each x, $x_0 \le x \le x_1$, we consider the line passing through $a(x)$ and

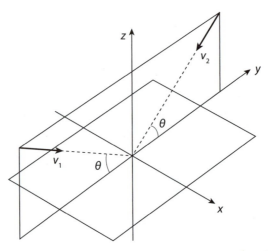

FIGURE 6. Two viewing directions parallel to the yz-plane.

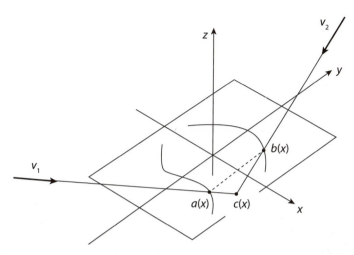

FIGURE 7. Space curve whose appearances coincide with two given plane curves.

parallel to v_1, and the line passing through $b(x)$ and parallel to v_2. These two lines are included in the same plane parallel to the yz-plane, and hence they have a point of intersection. Let this point be denoted by $c(x)$. Then, $c(x)$, $x_0 \leq x \leq x_1$, forms a space curve. The curve $c(x)$ coincides with $a(x)$ when it is seen in the direction v_1, and it coincides with $b(x)$ when it is seen in the direction v_2.

Finally, we choose a vertical line segment L and move it in such a way that it remains vertical, and the upper terminal point traces along the curve $c(x)$, $x_0 \leq x \leq x_1$. Let S be the surface swept by L. S is a surface with vertical rulers, and the vertical length of S is the same as the length of L. Therefore, when viewed, it appears to be a cylindrical surface with a constant height. In other words, we are likely to perceive it as a cylindrical surface whose upper and lower edges are obtained by cutting the surface with a plane perpendicular to the axis of the cylinder. This may be the result of the preference for rectangularity in the human vision system [11, 12]. As a result, the upper edge appears to be the plane curve $a(x)$ when seen in the direction v_1, and appears to be $b(x)$ when seen in the direction v_2. This method can be used to construct an ambiguous cylinder that has the two desired appearances $a(x)$ and $b(x)$ when it is seen from the special viewing directions v_1 and v_2, respectively.

The surface thus constructed is x-monotone and hence is not closed. If we want a closed cylinder, we can apply the above method twice, once for the upper half of the cylinder and once more for the lower half.

4. How to Make Topology-Disturbing Objects

We have reviewed a method for constructing a cylindrical surface whose upper edge has two desired appearances when it is viewed from two special directions. We can use this method to construct topology-disturbing objects, in the following way.

Consider the object shown in Figure 1; Figure 8 shows the shape of the sections of the object that we see. From the direct view, we perceive two nonintersecting rectangles as shown in (a), and from the mirror image, we perceive two intersecting rectangles as shown in (b). We decompose the shape in (b) into two nonintersecting closed curves, as shown in (c), where, for simplicity, the two closed curves are displaced so that they do not touch. We apply the method for ambiguous cylinders to the curves in (a) and (c). That is, we construct two ambiguous cylinders, one for the upper pair and one for the lower pair in Figure 8(a,c). Thus, we obtain two cylinders, each of which has the desired appearance.

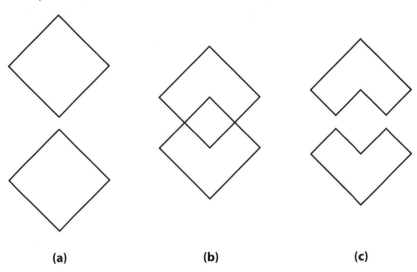

(a)   **(b)**   **(c)**

FIGURE 8. Desired appearance of two rectangular cylinders: (a) direct view; (b) view in the mirror; and (c) decomposition of (b) into two nonintersecting shapes.

The remaining problem is to combine the two cylinders so that they appear to be separated when viewed from the first direction v_1 and they appear to touch when viewed from the second view direction v_2; note that if we move the curves in Figure 8(c) so that they touch, they will appear as in (b), which will be perceived as intersecting. For this purpose, we place the two cylinders in different vertical positions, as shown in Figure 9. The two cylinders are placed apart, and then their vertical positions are adjusted so that they appear to touch when seen from one of the viewing directions, as shown by the broken line in the figure. We discuss later how to place the cylinders in more detail. From this method, we obtain a topology-disturbing object.

A general view of the object shown in Figure 1 is shown in Figure 10. As seen in this figure, the two cylinders are fixed at different heights; they are connected by additional material that is invisible from either of the two special viewing directions.

Note that a single image does not have depth information, and hence its interpretation as a 3D object is not unique. However, when we see the object and its mirror image from a special viewpoint shown in Figure 1, we usually perceive two separate cylinders and two intersecting cylinders. This perceptual phenomenon may be based on human

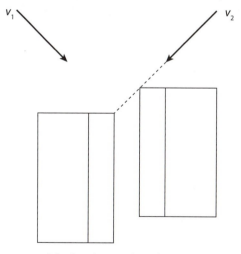

FIGURE 9. Adjustment of the heights so that they appear disconnected when viewed from one direction but they appear to intersect when viewed from the other direction.

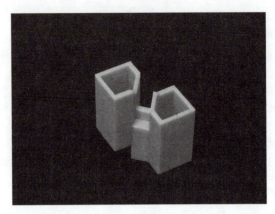

FIGURE 10. General view of the object shown in Figure 1.

preference and familiarity for canonical shapes, such as a circle and a square. Strength of similar preference has been observed and studied in many contexts, including figure-ground discrimination [8], depth perception [2], visual search [20], and line drawing interpretation [7]. This preference is an important psychological aspect of the topology-disturbing objects, but we postpone this issue for future work. The main issue of the present paper is to point out the geometric feasibility of this class of illusory objects.

The same approach can be used for the other examples of topology-disturbing objects that were presented in Section 2.

Figure 11 shows the perceived shapes of the object presented in Figure 2; Figure 11(a) shows the direct view, in which the two cylinders are separated, and Figure 11(b) shows the mirror image, in which the two cylinders intersect. We decompose the shape in (b) into two nonintersecting curves as shown in (c), and we then apply the above method to the curves in (a) and (c) to obtain two ambiguous cylinders. Finally, we adjust the distance between them and their heights in order to obtain the object shown in Figure 2. Figure 12 shows a general view of the resulting object.

Figure 13 shows the perceived shape of the object shown in Figure 3. Figure 13(a) is the direct view, and (b) is the mirror image. We decompose the shape in (b) into three nonintersecting curves, as shown in (c), and we then apply our method. A general view of the resulting object is shown in Figure 14.

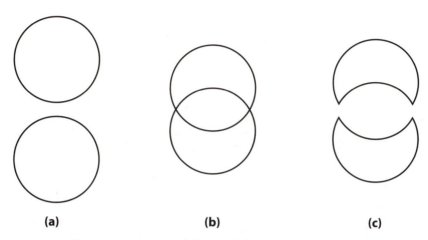

(a)   **(b)**   **(c)**

FIGURE 11. Perceived shapes of the object in Figure 2.

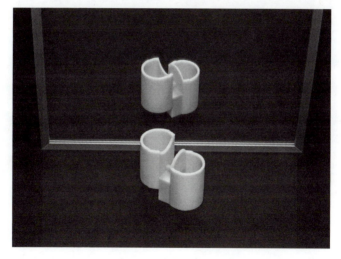

FIGURE 12. General view of the object shown in Figure 2.

For the object shown in Figure 4, we obtain the perceived shapes that are shown in Figure 15; (a) shows the direct view, (b) shows the mirror image, and (c) shows the decomposed nonintersecting curves, where the inner curve is shrunk slightly in order to clarify that they do not intersect. We apply our method to the images in (a) and (c) to obtain the object shown in Figure 4. A general view of that object is shown in Figure 16.

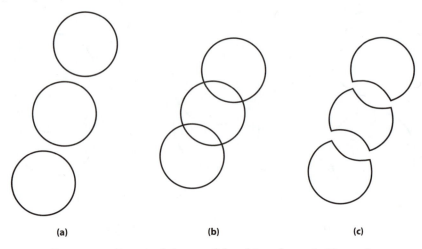

(a)   (b)   (c)

FIGURE 13. Perceived shapes of the object shown in Figure 3.

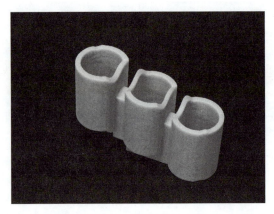

FIGURE 14. General view of the object shown in Figure 3.

For the object shown in Figure 5, we obtain the shape diagram shown in Figure 17; (a) shows the direct view, (b) shows the decomposed pair of nonintersecting curves, and (c) shows the mirror image. If we apply our method to (b) and (c), we obtain the object shown in Figure 5. A general view of the object is shown in Figure 18.

Three other examples of topology-disturbing objects are shown in Figures 19–21. In each figure, (a) shows the direct view and its image in a mirror, and (b) shows a general view of the object.

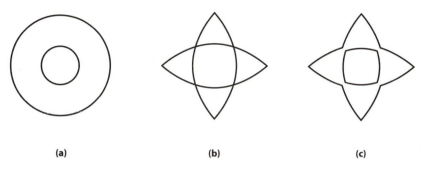

(a) (b) (c)

FIGURE 15. Perceived shapes of the object shown in Figure 4.

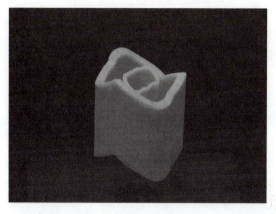

FIGURE 16. General view of the object shown in Figure 4.

Figure 19 shows an object composed of five cylinders. In the direct view, the three cylinders in the front row are touching, and the two cylinders in the back row are touching, but the two rows are separated. However, in the mirror image, all five cylinders intersect the adjacent cylinders.

Figure 20 shows an object composed of a cylinder and two parallel planes. In the direct view, both the planes cut through the cylinder, but in the mirror view, they are on opposite sides and separated from the cylinder.

Figure 21 shows another object. In the direct view, it consists of two intersecting lens-shaped cylinders, but in the image, they change to two pairs of touching but nonintersecting circular cylinders.

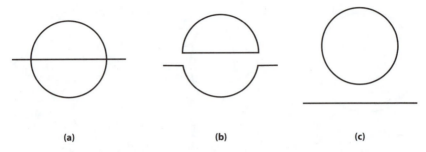

(a)   **(b)**   **(c)**

FIGURE 17. Perceived shapes of the object shown in Figure 5.

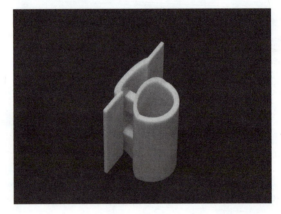

FIGURE 18. General view of the object shown in Figure 5.

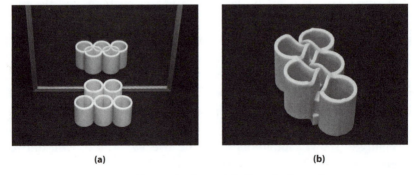

(a)   **(b)**

FIGURE 19. Five cylinders with disturbed topology.

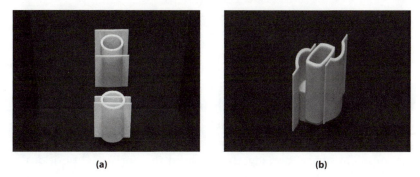

(a) **(b)**

FIGURE 20. Two parallel planes and a cylinder with disturbed topology.

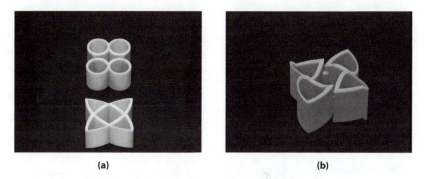

(a) **(b)**

FIGURE 21. Four cylinders with disturbed topology.

5. *Constructability Condition*

We have seen the construction process of topology-disturbing objects through examples. Next, we consider a general condition under which a real 3D object can be constructed from a given pair of appearances.

Let F be a line drawing composed of curves drawn on the xy-plane, where each curve is a nonself-intersecting closed curve, such as a circle or rectangle, or an open curve, such as a line segment. We assume that each curve in F is not self-intersecting even though different curves may intersect. We decompose the curves in F into x-monotone segments and represent them by equations $y = f_1(x)$, $y = f_2(x)$, . . . , $y = f_n(x)$ in such a way that

(1) two segments do not cross each other (although they may touch), and

(2) an upper segment has a smaller segment number than the lower segment, i.e., $f_i(x) \geq f_j(x)$ implies $i < j$.

For example, suppose that F is the line drawing composed of two circles and a line segment shown in Figure 22(a). This drawing can be decomposed into five x-monotone segments $f_1(x), f_2(x), \ldots, f_5(x)$ as shown in (b), where we slightly displaced the horizontal positions of curves so that the touching segments are separated for the convenience of understanding.

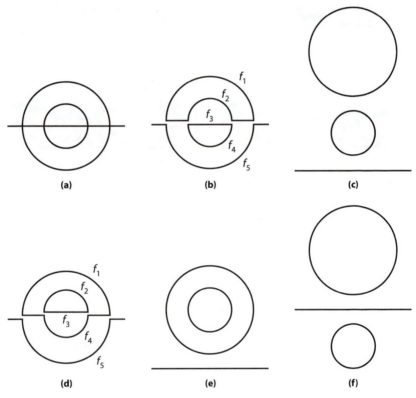

FIGURE 22. Line drawing and its decomposition into x-monotone segments: (a) first line drawing F; (b) decomposition of F; (c) second line drawing, which is consistent with (b); (d) another decomposition of F; (e) third line drawing, which is consistent with (d); and (f) fourth line drawing, which is not consistent with F.

Similarly, let G be another line drawing obtained from F by replacing each curve in the direction parallel to the y-axis, and $y = g_1(x)$, $y = g_2(x)$, . . . , $y = g_m(x)$ are x-monotone segments obtained by decomposing G, satisfying (1) and (2). We are interested in whether we can construct a cylindrical object whose appearances from two special viewpoints coincide with drawings F and G. We can prove the next theorem.

THEOREM 5.1: *We can construct a topology-disturbing object whose appearances coincide with drawings F and G if the following conditions are satisfied:*

(3) $m = n$, and
(4) for each i, $f_i(x)$ and $g_i(x)$ span the same x range; i.e., their leftmost points have the same x-coordinate, and their rightmost points also have the same x-coordinate.

A rough sketch of the proof is as follows. Suppose that conditions (3) and (4) are satisfied. Then we can construct the ambiguous cylinders corresponding to $f_i(x)$ and $g_i(x)$ for $i = 1, 2, . . . , n$. Let the resulting ambiguous cylinders be H_i, $i = 1, 2, . . . , n$. Next, place the vertical cylinders H_1, H_2, . . . , H_n so that their appearance coincides with drawing F when they are seen along the first view direction. This arrangement is always possible because the curves $f_1(x), f_2(x), . . . , f_n(x)$ do not cross each other (see condition (1)). If we see this collection of the cylinders along the second view direction, each cylinder has the desired appearance specified by the second line drawing G, but their mutual positions may not coincide with G. Therefore, as a final step, we translate the cylinders in the direction parallel to the first view direction so that the relative positions in the second appearance coincide with G. Note that the appearance in the first view direction is not changed by these translations because they are moved in that direction. Note also that these translations can be done without collision because their order in the y-direction is the same (see condition (2)). Thus, the theorem can be proved.

Let us go back to the example in Figure 22. Suppose that the second drawing G is given as in (c). This drawing can be decomposed into five x-monotone segments by cutting the circles at the leftmost and the rightmost points. This decomposition of G together with the decomposition of *F* into (b) satisfies the conditions of the theorem, and hence we can construct the associated topology-disturbing object.

The decomposition of the drawing in Figure 22(a) is not unique. Another decomposition is shown in (d). This decomposition and the unique decomposition of the line drawing in (e) satisfy the conditions in the theorem, and hence we can construct a topology-disturbing object corresponding to drawings (a) and (e).

Figure 22(f) shows still another drawing composed of the same two circles and the line segment. However, we cannot decompose drawing (a) so that it is consistent with (f) in terms of condition (4). Hence, we cannot construct a topology-disturbing object associated with drawings (a) and (f).

6. Concluding Remarks

We have presented a class of illusory objects, called topology-disturbing objects, which, when viewed from two special directions, appear to have different topologies; this change of view gives us the impression that they are physically impossible. Therefore, this can be regarded as a new class of impossible objects. Two viewing directions can be realized simultaneously by using a mirror, and hence, these objects can be displayed effectively (such as for an exhibition) by using a mirror.

These objects may be used in many contexts. In vision science, they offer new material for research on seeing; for example, in order to understand human vision, we must clarify why we easily perceive topological inconsistencies. In science education, these objects could be used to stimulate children to start thinking about visual perception. At least upon one's first encounter with topology-disturbing objects, they are surprising and mysterious, and thus their unusual visual effects might be used in arts and entertainment. One aspect of our future work is to investigate the possibilities in these directions.

Acknowledgments

The author expresses his sincere thanks to the anonymous reviewers for their valuable comments, by which he was able to improve the manuscript.

Disclosure Statement

No potential conflict of interest was reported by the author.

Funding

Grant-in-Aid for Challenging Exploratory Research [Grant number 15K2067] and for Basic Research (A) [Grant number 16H01728] from MEXT.

References

[1] F. H. Bool, J. R. Kist, J. L. Locher, and F. Wierda, M. C. Escher: His Life and Complete Graphic Work with a Fully Illustrated Catalogue, Harry N. Abrams, New York, 1992.

[2] I. Bülthoff, H. Bülthoff, and P. Sinha, Top-down influences on stereoscopic depth-perception, *Nature* 1(3) (1998), pp. 254–257.

[3] J. J. Dobias, T. V. Papathomas, and V. M. Vlajnic, Convexity bias and perspective cues in the reverse-perspective illusion, *i-Percept* 7 (2016), pp. 1–7.

[4] B. Ernst, Impossible World, Taschen GmbH, Cologne, Germany, 2006.

[5] M. C. Escher, M. C. Escher: The Graphic Work, 25th ed., Taschen America, New York, 2008.

[6] R. L. Gregory, The Intelligent Eye, Weidenfeld and Nicolson, London, 1970.

[7] P. Mamassian and M. S. Landy, Observer biases in the 3D interpretation of line drawings, *Vision Res.* 38 (1998), pp. 2817–2832.

[8] R. A. Nelson and S. E. Palmer, Familiar shapes attract attention in figure-ground display, *Percept Psychophys.* 69 (2007), pp. 382–392.

[9] T. V. Papathomas, Art pieces that "move" in our minds: An explanation of illusory motion based on depth reversal, *Spatial Vision* 21 (2007), pp. 79–95.

[10] L. S. Penrose and R. Penrose, Impossible objects: A special type of visual illusion, *British J. Psychol.* 49 (1958), pp. 31–33.

[11] D. N. Perkins, Visual discrimination between rectangular and nonrectangular parallelo-pipeds, *Percept Psychophys.* 12 (1972), pp. 396–400.

[12] D. N. Perkins, Compensating for distortion in viewing pictures obliquely, *Percept Psychophys.* 14 (1973), pp. 13–18.

[13] V. S. Ramachandran, Perception of shape from shading, *Nature* 331 (1988), pp. 163–166.

[14] K. Sugihara, Impossible motion: Magnet-like slopes. 1st prize illusion of the 6th best illusion of the year contest (2010). Available at https://www.youtube.com/watch?v=fYa0y4ETFVo.

[15] K. Sugihara, Design of solids for antigravity motion illusion, *Comput. Geom. Theory Appl.* 47 (2014), pp. 675–682.

[16] K. Sugihara, Ambiguous cylinders: A new class of impossible objects, *CADDM.* 25(3) (2015), pp. 19–25.

[17] K. Sugihara, Height reversal generated by rotation around a vertical axis, *J. Math. Psychol.* 68–69 (2015), pp. 7–12.

[18] K. Sugihara, Impossible Objects 10: Ambiguous Cylinders 4, YouTube video. Available at https://youtu.be/bylrPtYBBQI.

[19] N. J. Wade and P. Hughes, Fooling the eyes: Trompe l'Oeil and reverse perspective, *Perception* 28 (1999), pp. 1115–1119.

[20] Q. Wang and P. Cavanagh, Familiarity and pop-out in visual search, *Percept Psychophys.* 56 (1994), pp. 495–500.

Appendix A.
Unfolded Surfaces of a Topology-Disturbing Object

The topology-disturbing objects shown in this paper were made by a 3D printer. Since they are in general composed of complicated curved surfaces, it is difficult to make them by hand. However, the object shown in Figure 1 is an exception. It consists of planar faces, and hence, we can use paper crafting to construct it.

Figure A.1 shows a diagram of the unfolded surfaces of the object. It consists of three components: A, B, and C. Components A and B correspond to the two rectangular cylinders, and component C is used to

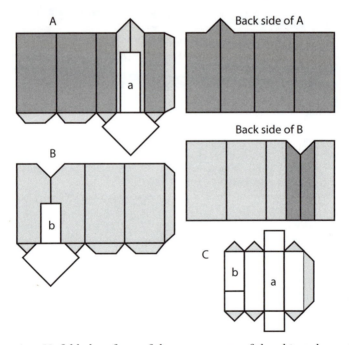

FIGURE A.1. Unfolded surfaces of the components of the object shown in Figure 1.

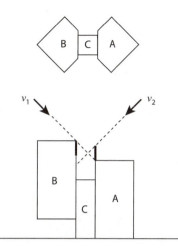

FIGURE A.2. Top and side views of the object.

connect them. The light gray triangular and trapezoidal areas indicate where they should be glued together. For components A and B, both the front and back sides are shown. Copy them onto stiff paper, cut out and fold as indicated (solid interior lines). The white areas indicated by a and b are glued to the matching sections of C.

Figure A.2 shows the top view (upper part) and the side view (lower part) of the constructed object. As shown in this figure, component B should be placed slightly higher than component A. The viewing angle for this object is 45 degrees; that is, if we view this object from above at an angle of 45 degrees from the horizontal, as shown by v_1 in Figure A.2, the corners of the two cylinders will coincide and thus appear to intersect. If viewed from the opposite side, v_2, the two cylinders appear to be disconnected, because they are at different heights.

We encourage the readers to construct this object for themselves in order to better understand the actual shape of the object and the way in which we perceive this illusion.

Mathematicians Explore Mirror Link between Two Geometric Worlds

Kevin Hartnett

Twenty-seven years ago, a group of physicists made an accidental discovery that flipped mathematics on its head. The physicists were trying to work out the details of string theory when they observed a strange correspondence: Numbers emerging from one kind of geometric world matched exactly with very different kinds of numbers from a very different kind of geometric world.

To physicists, the correspondence was interesting. To mathematicians, it was preposterous. They'd been studying these two geometric settings in isolation from each other for decades. To claim that they were intimately related seemed as unlikely as asserting that at the moment an astronaut jumps on the moon, some hidden connection causes his sister to jump back on Earth.

"It looked totally outrageous," said David Morrison, a mathematician at the University of California, Santa Barbara, and one of the first mathematicians to investigate the matching numbers.

Nearly three decades later, incredulity has long since given way to revelation. The geometric relationship that the physicists first observed is the subject of one of the most flourishing fields in contemporary mathematics. The field is called *mirror symmetry*, in reference to the fact that these two seemingly distant mathematical universes appear somehow to reflect each other exactly. And since the observation of that first correspondence—a set of numbers on one side that matched a set of numbers on the other—mathematicians have found many more instances of an elaborate mirroring relationship: Not only do the astronaut and his sister jump together, they wave their hands and dream in unison, too.

Recently, the study of mirror symmetry has taken a new turn. After years of discovering more examples of the same underlying phenomenon, mathematicians are closing in on an explanation for why the phenomenon happens at all.

"We're getting to the point where we've found the ground. There's a landing in sight," said Denis Auroux, a mathematician at the University of California, Berkeley.

The effort to come up with a fundamental explanation for mirror symmetry is being advanced by several groups of mathematicians. They are closing in on proofs of the central conjectures in the field. Their work is like uncovering a form of geometric DNA—a shared code that explains how two radically different geometric worlds could possibly hold traits in common.

Discovering the Mirror

What would eventually become the field of mirror symmetry began when physicists went looking for some extra dimensions. As far back as the late 1960s, physicists had tried to explain the existence of fundamental particles—electrons, photons, quarks—in terms of minuscule vibrating strings. By the 1980s, physicists understood that in order to make string theory work, the strings would have to exist in 10 dimensions—six more than the four-dimensional space-time we can observe. They proposed that what went on in those six unseen dimensions determined the observable properties of our physical world.

"You might have this small space that you can't see or measure directly, but some aspects of the geometry of that space might influence real-world physics," said Mark Gross, a mathematician at the University of Cambridge.

Eventually, they came up with potential descriptions of the six dimensions. Before getting to them, though, it's worth thinking for a second about what it means for a space to have a geometry.

Consider a beehive and a skyscraper. Both are three-dimensional structures, but each has a very different geometry: Their layouts are different, the curvature of their exteriors is different, their interior angles are different. Similarly, string theorists came up with very different ways to imagine the missing six dimensions.

One method arose in the mathematical field of algebraic geometry. Here, mathematicians study polynomial equations—for example, $x^2 + y^2 = 1$—by graphing their solutions (a circle, in this case). More-complicated equations can form elaborate geometric spaces. Mathematicians explore the properties of those spaces to better understand the original equations. Because mathematicians often use complex numbers, these spaces are commonly referred to as "complex" manifolds (or shapes).

The other type of geometric space was first constructed by thinking about physical systems such as orbiting planets. The coordinate values of each point in this kind of geometric space might specify, for example, a planet's location and momentum. If you take all possible positions of a planet together with all possible momenta, you get the "phase space" of the planet—a geometric space whose points provide a complete description of the planet's motion. This space has a *symplectic* structure that encodes the physical laws governing the planet's motion.

Symplectic and complex geometries are as different from one another as beeswax and steel. They make very different kinds of spaces. Complex shapes have a very rigid structure. Think again of the circle. If you wiggle it even a little, it's no longer a circle. It's an entirely distinct shape that can't be described by a polynomial equation. Symplectic geometry is much floppier. There, a circle and a circle with a little wiggle in it are almost the same.

"Algebraic geometry is a more rigid world, whereas symplectic geometry is more flexible," said Nick Sheridan, a research fellow at Cambridge. "That's one reason they're such different worlds, and it's so surprising they end up being equivalent in a deep sense."

In the late 1980s, string theorists came up with two ways to describe the missing six dimensions: one derived from symplectic geometry, the other from complex geometry. They demonstrated that either type of space was consistent with the four-dimensional world they were trying to explain. Such a pairing is called a *duality*: Either one works, and there's no test you could use to distinguish between them.

Physicists then began to explore just how far the duality extended. As they did so, they uncovered connections between the two kinds of spaces that grabbed the attention of mathematicians.

In 1991, a team of four physicists—Philip Candelas, Xenia de la Ossa, Paul Green, and Linda Parkes—performed a calculation on the complex side and generated numbers that they used to make predictions about corresponding numbers on the symplectic side. The prediction had to do with the number of different types of curves that could be drawn in the six-dimensional symplectic space. Mathematicians had long struggled to count these curves. They had never considered that these counts of curves had anything to do with the calculations on complex spaces that physicists were now using in order to make their predictions.

The result was so far-fetched that at first, mathematicians didn't know what to make of it. But then, in the months following a hastily convened meeting of physicists and mathematicians in Berkeley, California, in May 1991, the connection became irrefutable. "Eventually mathematicians worked on verifying the physicists' predictions and realized this correspondence between these two worlds was a real thing that had gone unnoticed by mathematicians who had been studying the two sides of this mirror for centuries," said Sheridan.

The discovery of this mirror duality meant that in short order, mathematicians studying these two kinds of geometric spaces had twice the number of tools at their disposal: Now they could use techniques from algebraic geometry to answer questions in symplectic geometry, and vice versa. They threw themselves into the work of exploiting the connection.

Breaking Up Is Hard to Do

At the same time, mathematicians and physicists set out to identify a common cause, or underlying geometric explanation, for the mirroring phenomenon. In the same way that we can now explain similarities between very different organisms through elements of a shared genetic code, mathematicians attempted to explain mirror symmetry by breaking down symplectic and complex manifolds into a shared set of basic elements called *torus fibers*.

A torus is a shape with a hole in the middle. An ordinary circle is a one-dimensional torus, and the surface of a donut is a two-dimensional torus. A torus can be of any number of dimensions. Glue lots of lower dimensional tori together in just the right way, and you can build a higher dimensional shape out of them.

To take a simple example, picture the surface of the Earth. It is a two-dimensional sphere. You could also think of it as being made from many one-dimensional circles (like many lines of latitude) glued together. All these circles stuck together are a *torus fibration* of the sphere—the individual fibers woven together into a greater whole.

Torus fibrations are useful in a few ways. One is that they give mathematicians a simpler way to think of complicated spaces. Just as you can construct a torus fibration of a two-dimensional sphere, you can construct a torus fibration of the six-dimensional symplectic and complex spaces that feature in mirror symmetry. Instead of circles, the fibers of those spaces are three-dimensional tori. And whereas a six-dimensional symplectic manifold is impossible to visualize, a three-dimensional torus is almost tangible. "That's already a big help," said Sheridan.

A torus fibration is useful in another way: It reduces one mirror space to a set of building blocks that you could use to build the other. In other words, you can't necessarily understand a dog by looking at a duck, but if you break each animal into its raw genetic code, you can look for similarities that might make it seem less surprising that both organisms have eyes.

Here, in a simplified view, is how to convert a symplectic space into its complex mirror. First, perform a torus fibration on the symplectic space. You'll get a lot of tori. Each torus has a radius (just like a circle—a one-dimensional torus—has a radius). Next, take the reciprocal of the radius of each torus. (So, a torus of radius 4 in your symplectic space becomes a torus of radius 1/4 in the complex mirror.) Then use these new tori, with reciprocal radii, to build a new space.

In 1996, Andrew Strominger, Shing-Tung Yau, and Eric Zaslow proposed this method as a general approach for converting any symplectic space into its complex mirror. The proposal that it's always possible to use a torus fibration to move from one side of the mirror to the other is called the SYZ conjecture, after its originators. Proving it has become one of the foundational questions in mirror symmetry (along with the homological mirror symmetry conjecture, proposed by Maxim Kontsevich in 1994).

The SYZ conjecture is hard to prove because, in practice, this procedure of creating a torus fibration and then taking reciprocals of the radii is not easy to do. To see why, return to the example of the surface of the Earth. At first it seems easy to stripe it with circles, but at the poles, your circles have a radius of zero. And the reciprocal of zero is infinity. "If your radius equals zero, you've got a bit of a problem," said Sheridan.

This same difficulty crops up in a more pronounced way when you're trying to create a torus fibration of a six-dimensional symplectic space. There, you might have infinitely many torus fibers where part of the fiber is pinched down to a point—points with a radius of zero. Mathematicians are still trying to figure out how to work with such fibers. "This torus fibration is really the great difficulty of mirror symmetry," said Tony Pantev, a mathematician at the University of Pennsylvania.

Put another way, the SYZ conjecture says that a torus fibration is the key link between symplectic and complex spaces, but in many cases, mathematicians don't know how to perform the translation procedure that the conjecture prescribes.

Long-Hidden Connections

Over the past 27 years, mathematicians have found hundreds of millions of examples of mirror pairs: This symplectic manifold is in a mirror relationship with that complex manifold. But when it comes to understanding why a phenomenon occurs, quantity doesn't matter. You could assemble an ark's worth of mammals without coming any closer to understanding where hair comes from.

"We have huge numbers of examples, like 400 million examples. It's not that there's a lack of examples, but nevertheless it's still specific cases that don't give much of a hint as to why the whole story works," said Mark Gross.

Mathematicians would like to find a general method of construction—a process by which you could hand them any symplectic manifold and they could hand you back its mirror. And now they believe that they're getting close to having it. "We're moving past the case-by-case understanding of the phenomenon," said Auroux. "We're trying to prove that it works in as much generality as we can."

Mathematicians are progressing along several interrelated fronts. After decades building up the field of mirror symmetry, they're close to understanding the main reasons the field works at all.

"I think it will be done in a reasonable time," said Kontsevich, a mathematician at the Institute of Advanced Scientific Studies (IHES) in France and a leader in the field. "I think it will be proven really soon."

One active area of research creates an end run around the SYZ conjecture. It attempts to port geometric information from the symplectic side to the complex side without a complete torus fibration. In 2016, Gross and his longtime collaborator Bernd Siebert of the University of Hamburg posted a general-purpose method for doing so. They are now finishing a proof to establish that the method works for all mirror spaces. "The proof has now been completely written down, but it's a mess," said Gross, who said that he and Siebert hope to complete it by the end of the year.

Another major open line of research seeks to establish that, assuming you have a torus fibration, which gives you mirror spaces, then all the most important relationships of mirror symmetry fall out from there. The research program is called *family Floer theory* and is being developed by Mohammed Abouzaid, a mathematician at Columbia University. In March 2017, Abouzaid posted a paper that proved that this chain of logic holds for certain types of mirror pairs, but not yet all of them.

And, finally, there is work that circles back to where the field began. A trio of mathematicians—Sheridan, Sheel Ganatra, and Timothy Perutz—is building on seminal ideas introduced in the 1990s by Kontsevich related to his homological mirror symmetry conjecture.

Cumulatively, these three initiatives would provide a potentially complete encapsulation of the mirror phenomenon. "I think we're getting to the point where all the big 'why' questions are close to being understood," said Auroux.

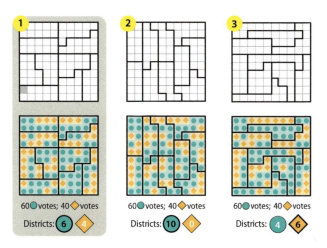

60 ⬤ votes; 40 ◆ votes 　60 ⬤ votes; 40 ◆ votes 　60 ⬤ votes; 40 ◆ votes

Districts: (6) ◆ 4 　　Districts: (10) ◆ 0 　　Districts: (4) ◆ 6

FIGURE 1 from "Geometry vs. Gerrymandering" (Duchin)

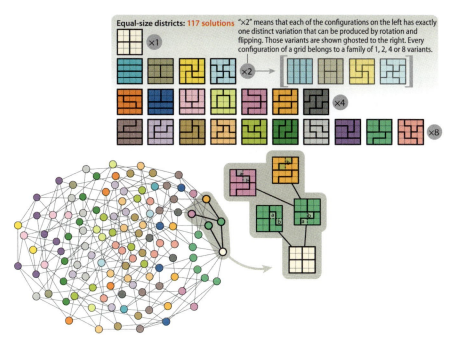

Equal-size districts: 117 solutions "×2" means that each of the configurations on the left has exactly one distinct variation that can be produced by rotation and flipping. Those variants are shown ghosted to the right. Every configuration of a grid belongs to a family of 1, 2, 4 or 8 variants.

FIGURE 3 from "Geometry vs. Gerrymandering" (Duchin)

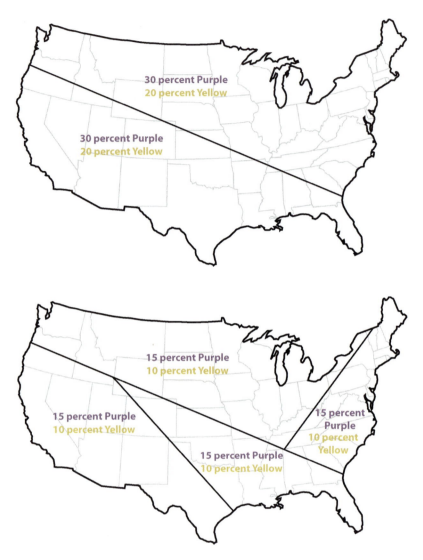

Figure 2 from "Slicing Sandwiches, States, and Solar Systems" (Hill)

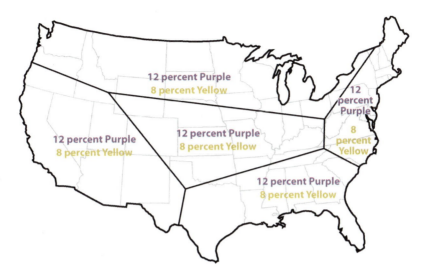

Figure 3 from "Slicing Sandwiches, States, and Solar Systems" (Hill)

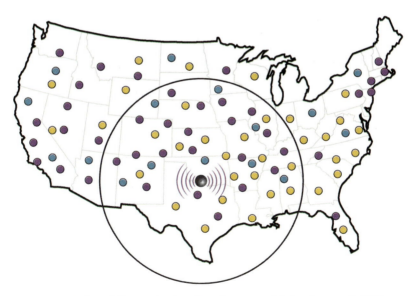

Figure 5 from "Slicing Sandwiches, States, and Solar Systems" (Hill)

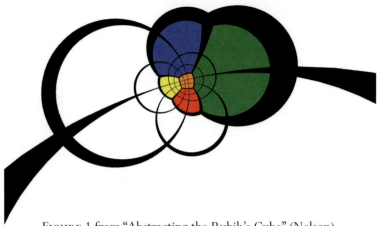

FIGURE 1 from "Abstracting the Rubik's Cube" (Nelson)

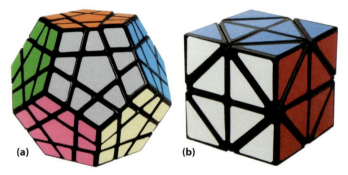

(a)   (b)

FIGURE 2 from "Abstracting the Rubik's Cube" (Nelson)

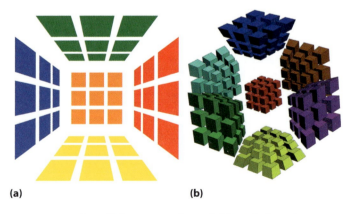

(a)   (b)

FIGURE 3 from "Abstracting the Rubik's Cube" (Nelson)

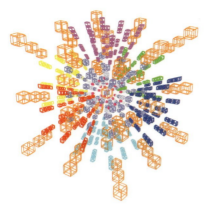

FIGURE 4 from "Abstracting the Rubik's Cube" (Nelson)

FIGURE 5 from "Abstracting the Rubik's Cube" (Nelson)

(a)

(b)

FIGURE 6 from "Abstracting the Rubik's Cube" (Nelson)

FIGURE 7 from "Abstracting the Rubik's Cube" (Nelson)

Figure 8 from "Abstracting the Rubik's Cube" (Nelson)

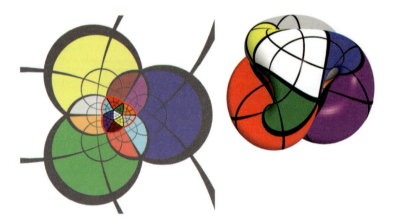

Figure 9 from "Abstracting the Rubik's Cube" (Nelson)

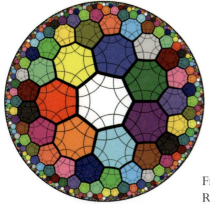

Figure 10 from "Abstracting the Rubik's Cube" (Nelson)

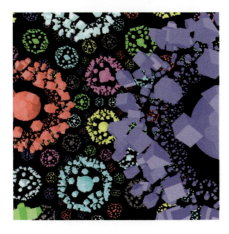

Figure 11 from "Abstracting the Rubik's Cube" (Nelson)

Figure 12 from "Abstracting the Rubik's Cube" (Nelson)

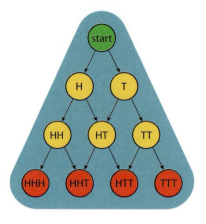

FIGURE 1 from "Professor Engel's Marvelously Improbable Machines" (Propp)

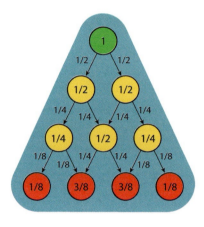

FIGURE 2 from "Professor Engel's Marvelously Improbable Machines" (Propp)

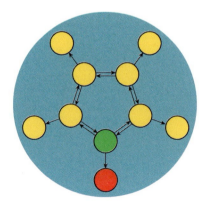

FIGURE 3 from "Professor Engel's Marvelously Improbable Machines" (Propp)

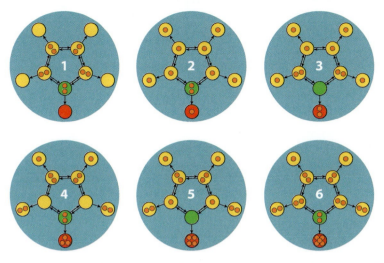

FIGURE 4 from "Professor Engel's Marvelously Improbable Machines"
(Propp)

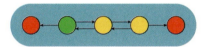

FIGURE 5 from "Professor Engel's Marvelously Improbable Machines"
(Propp)

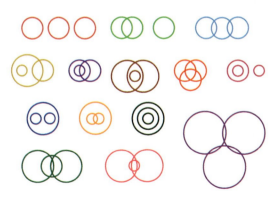

FIGURE 6 from "The On-Line Encyclopedia of Integer Sequences" (Sloane).
Courtesy of Jessica Gonzalez.

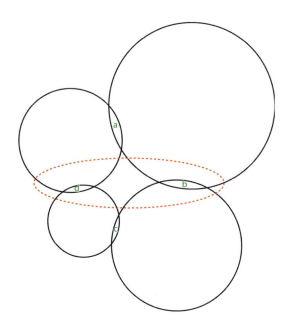

FIGURE 9 from "The On-Line Encyclopedia of Integer Sequences" (Sloane).
Courtesy of Jonathan Wild.

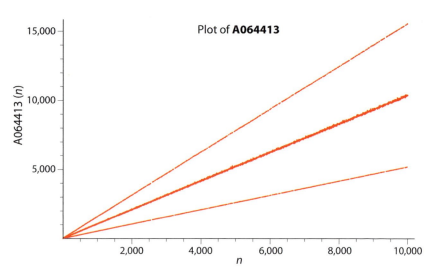

FIGURE 10 from "The On-Line Encyclopedia of Integer Sequences" (Sloane).
Courtesy of Michael De Vlieger

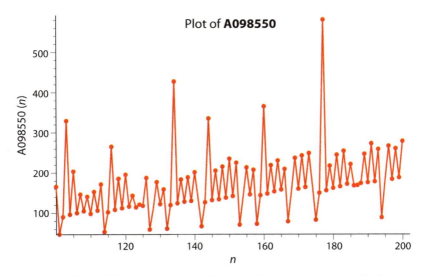

FIGURE 11 from "The On-Line Encyclopedia of Integer Sequences" (Sloane).
Courtesy of Michael De Vlieger

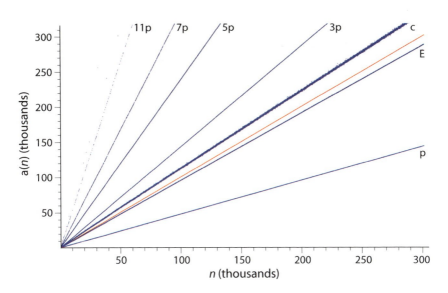

FIGURE 12 from "The On-Line Encyclopedia of Integer Sequences" (Sloane).
Courtesy of Hans Havermann

Figure 14 from "The On-Line Encyclopedia of Integer Sequences" (Sloane).
Courtesy of Kerry Mitchell

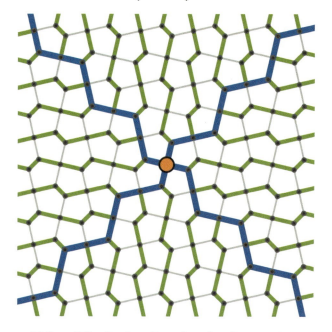

Figure 18 from "The On-Line Encyclopedia of Integer Sequences"
(Sloane). Courtesy of Chaim Goodman-Strauss

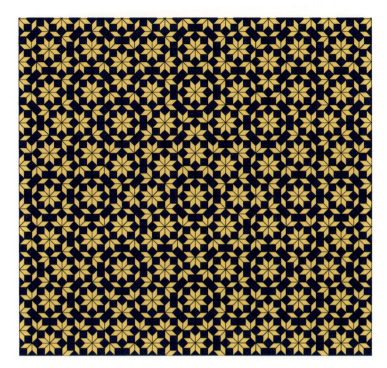

Figure 19 from "The On-Line Encyclopedia of Integer Sequences" (Sloane). From *The Tilings Encyclopedia*, https://tilings.math.unibielefeld.de. Used under the Creative Commons Attribution-NonCommercial-ShareAlike 2.0 Generic License

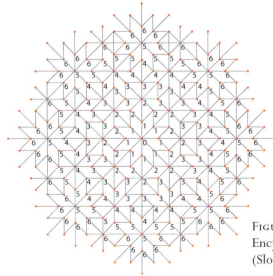

Figure 20 from "The On-Line Encyclopedia of Integer Sequences" (Sloane). Courtesy of Rémy Sigrist

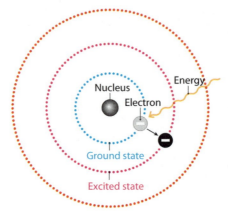

FIGURE 1a from "The Un(solv)able Problem" (Cubitt/Pérez-García/Wolf)

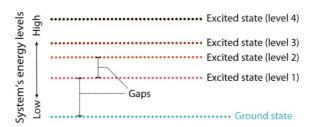

FIGURE 1b from "The Un(solv)able Problem" (Cubitt/Pérez-García/Wolf)

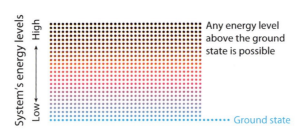

FIGURE 1c from "The Un(solv)able Problem" (Cubitt/Pérez-García/Wolf)

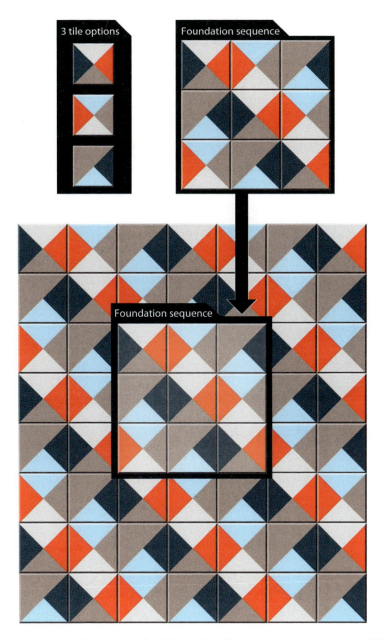

FIGURE 3 from "The Un(solv)able Problem" (Cubitt/Pérez-García/Wolf)

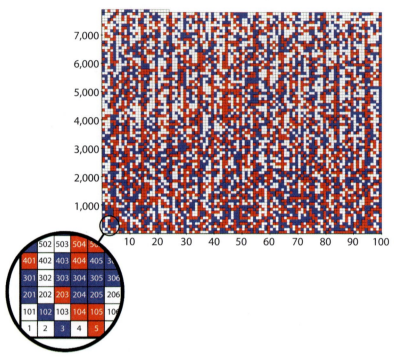

FIGURE 1 from "The Mechanization of Mathematics" (Avigad).
Courtesy of Marijn J. Heule

Professor Engel's Marvelously Improbable Machines

James Propp

When the path from simple question to simple answer leads you through swamps of computation, you can either accept that some amount of tromping through swamps is unavoidable, or you can think harder and try to find a different route. This is a story of someone who thought harder.

In the 1970s, the German mathematician Arthur Engel was teaching probability theory and other topics to middle and high school students in Illinois. He asked them questions like, "How many times should you expect to roll a fair die until it shows a three?" Such simple questions tend to have simple answers: whole numbers or fractions with small numerators and denominators. Engel taught his students to solve them using fraction arithmetic (in simpler cases) or small systems of linear equations (for more complicated problems).

In 1974, Engel was teaching fourth graders who were struggling with basic topics like adding and multiplying fractions. He wanted to teach them probability. But how? They could approach probability questions about spinners, dice, and coins experimentally, but if they couldn't follow up by computing exact answers, wouldn't they be missing a key part of the learning experience?

Engel got around the impasse by teaching them how to build and operate devices that could efficiently compute exact answers to probability problems. All they needed was paper, pencils, and small objects, such as beads or buttons.

An Easy Problem

If we toss a fair coin three times, what's the probability we get heads twice and tails once (in no particular order)?

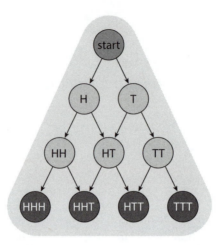

FIGURE 1. Possibilities for three coin tosses. See also color images.

Figure 1 shows the possible number of heads and tails we can see after zero, one, two, or three tosses. The second node at the bottom of the diagram is HHT, which means two heads and one tail. We also have arrows that lead from a node in any row (other than the bottom row) to two nodes in the next row, corresponding to the fact that at each stage in the coin-tossing process, there are two equally likely possibilities.

The process of tossing a coin three times can be shadowed by placing a token on the top node and sliding it left or right when the coin shows heads or tails, respectively. The token's path becomes the object of study rather than the sequence of coin flips; we call the path a *random walk*. The rule for random walking is, when the token is at a node with outgoing arrows, we choose an arrow at random and slide the token along it.

To analyze this random walk, we work our way from top to bottom, assigning a value to each node and each arrow. The number assigned to a node is the probability our token will visit the node, and the number on an arrow shows the probability our token will slide along that arrow. The probability on the top node is 1. If we've computed the probability p at a node, the two outgoing arrows should each be assigned the value p times $1/2$, or $p/2$. If we've computed the probabilities p and q for two arrows that point to the same node, that node gets the value $p + q$ (Figure 2).

Following this procedure, we conclude that the probabilities of seeing all heads, two heads and one tail, one head and two tails, and all tails, are $1/8$, $3/8$, $3/8$, and $1/8$, respectively.

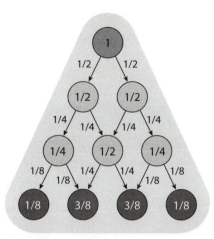

FIGURE 2. Probabilities for each node and arrow. See also color images.

But there's a way to compute these probabilities that avoids using fractions until the very end. Engel realized that, properly deployed, the diagram can be a kind of computer. We'll use small objects called "chips" that we can slide around but that won't roll away. Put eight chips on the start node (we'll see shortly where the number eight comes from). When there are k chips at a node, we equitably move $k/2$ to the left and $k/2$ to the right. We call this *chip firing*. Eight chips in the top row become $4 + 4$ chips in the next row, which become $2 + 4 + 2$ chips in the next, and $1 + 3 + 3 + 1$ in the final row. The number of chips in the second node in the last row (3) divided by the total number of chips (8) yields $3/8$.

Why Start with Eight?

How did we know to put eight chips into the machine? If we'd started with a number of chips that wasn't a multiple of eight, at some point it would have been impossible to divide them equitably between the outgoing arrows. If we had started with seven chips, we would have hit a snag right away: We couldn't have divided the pile into two equal halves. The best we could have done would have been to send three chips left and three right, but one chip would have been stuck on the top node.

Engel's first big insight was that this stuckness isn't a fatal flaw for our procedure. We can leave a stuck chip where it is for a while, trusting that sometime later another chip will arrive to free it. In our seven-chip example, if there's one chip stuck at the top, adding one more chip (the eighth) will result in two chips at the top node, and then those two chips can fire.

This leads to a chip-firing algorithm for solving the coin-toss problem that doesn't involve knowing the magic number eight in advance: Add a chip at the top, fire all possible nodes, add another chip at the top, fire all possible nodes, and so on, until no chips are stuck—that is, until there are no chips in the top three rows. This happens after eight chips have been added, but we didn't need to know this in advance; we just kept going until, magically, the first three rows of the board got cleared out.

The reader is strongly encouraged to pause and carry out this process—draw the diagram on paper and use some chips to carry out firing. I used chocolate chips.

Keep in mind that Engel's procedure is not a random simulation, but rather a distinctly nonrandom average-case simulation. We call it a *quasirandom simulation*. If we were to perform eight experiments in which we toss a fair coin three times, we might expect on average four of the experiments to have the coin show heads on the first toss, but that's only the *average* case, and in fact, the chance our eight experiments will split up in this equitable fashion is only about 27%. Real-world randomness is, as Robert Abelson aptly put it, "lumpy." Quasirandom simulation smooths out the lumps, which is good if we're interested in average-case behavior, but bad if we're interested in typical behavior.

A Harder Problem

The coin-toss problem had a fixed duration. What about random processes that can in principle last any finite number of steps? Let's look at Bruce Torrence's "Who keeps the money you found on the floor?" puzzle, which was the September 9, 2016, Riddler on the Five Thirty Eight website (http://53eig.ht/2MsUPtL): Five players sit around a table with a $100 bill in front of one player. The game is played in turns. On each turn, one of three equally likely things can happen: The bill moves one position left, it moves one position right, or the game ends with the money holder keeping the cash. What is the probability the starting player wins the money?

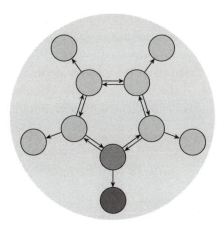

FIGURE 3. The diagram for Torrence's problem. See also color images.

We can analyze this problem using random walks. Figure 3 shows a diagram with five *interior nodes*, five *boundary nodes*, and three outgoing arrows from each interior node. When the randomly walking token moves from an interior node, it has an equal chance (1/3) of following each arrow. Boundary nodes have no outgoing arrows; when the token arrives at such a node, it stays there forever. We want to know: What is the chance that a randomly walking token placed at the green starting node ends at the red node?

One thing that makes this problem hard is that we can't bound the length of the game ahead of time. It's unlikely it will last a million moves (say), but this possibility has a positive chance of happening, so if we pretend it can't happen, we'll get the wrong answer. At the same time, the probability the game will go on forever is zero, so sooner or later, the token will arrive at one of the boundary nodes.

One might think the probability that the token ends at each boundary node is 1/5, but that can't be right; after all, the probability it ends up at the red node on the first move is 1/3.

We can try solving Torrence's problem using chip firing, as we did with the coin-toss problem: Begin with an empty board. Add chips to the start node one at a time. When any interior node has three or more chips, send one chip to each neighbor. Continue firing until every interior node has at most two chips on it (that is, until the loading of the chips at the interior nodes is *stable*). Every time we add a new chip at the green node, do as many chip-firing operations as are required to restore stability. Continue until there are no chips at any interior node.

However, this algorithm does not work! When an interior node fires, it sends two chips to other interior nodes. So, although many chips find their way to the boundary, there will always be some chips in the interior. If our rule for stopping the computation is to "wait until the interior nodes are clear," our procedure will go on forever!

Engel's Key Insight

By playing around with many problems of this kind, Engel hit on his second key idea. Instead of beginning with an *empty* loading of the interior nodes, begin with as *full* a loading as possible, subject to the constraint that we don't want any firing to happen at the beginning. In this setting, an interior node is a node that has one or more outgoing arrows. In the case of the 10-node machine we built for Torrence's problem, we put two chips at each of the interior nodes (which is the greatest possible number of chips if we don't want any firing to happen). This is called the *maximal stable loading*, or the *critical loading*, of the interior nodes. Our stopping rule is that we end the computation when the critical loading recurs. When it does, we can read off the answer by counting the chips that have been absorbed by the various nodes on the boundary.

Figure 4 shows the chip firing process applied to Torrence's problem (see also the video youtu.be/Ap5NslpYikU). It begins in (1) with the critical loading of the nodes. When we add a chip to the green node, it fires, sending one chip to the red node and one to each neighbor. Then the neighboring nodes fire, and subsequently, the last two interior nodes fire. We end with the chips as in (2). When we add a chip to the green node, it fires. That yields the configuration in (3). At this point, the green node is empty, so it won't fire until we add three chips. When it does, it yields the configuration in (4). Adding one chip produces (5). Adding three chips yields a firing and an empty green node. When we add two chips to the green node we return to the critical loading state in (6). Finally the punch line: The boundary nodes absorbed 11 chips, 5 of which are in the red node; so the answer to Torrence's question is 5/11.

This method works for every picture we can draw with nodes and arrows, not just for the one from Torrence's problem.

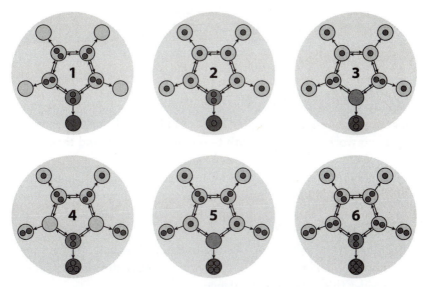

FIGURE 4. Chip firing solution for Torrence's problem. See also color images.

Engel called his invention the "probabilistic abacus," and later, the "stochastic abacus," but it is more accurate to say he came up with a method for designing and operating a host of abacuses, each tailored to the problem at hand.

An Engel abacus is digital in the sense that, like a digital computer, it routes discrete "impulses" (chips) along "wires" (arrows). But it is "analog" in the sense that it is based on an analogy between a random walk and a quasirandom walk. Indeed, one way to prove that Engel's abacuses give correct answers is to set up the analogy explicitly, and to show that the "probability flow" associated with a random walk satisfies the same equations as the flow of chips in the network. Details appear in Engel's article "Why does the probabilistic abacus work?" (*Educ. Stud. Math.* 7, no. 1/2 [July 1976]: 59–69).

Ruinous Gamblers

We don't want to end without giving readers a chance to play with Engel's ideas on their own. So, let's introduce a famous random process and invite readers to study it via Engel's method. This is the gambler's ruin problem. Rather than motivate it with the usual oversimplified

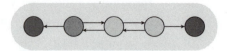

FIGURE 5. The gambler's ruin network. See also color images.

model of gambling, we present it as a random walk problem on the simple network in Figure 5 with two boundary nodes and three interior nodes.

A randomly walking token at an interior node moves left or right, each with probability 1/2. A token that arrives at a boundary node stays there forever. It's reasonable to suppose (and not too hard to prove) that the token will eventually become absorbed at a boundary node. But which one? If the token starts in the middle, it's equally likely to end up at either node, but what if the token starts off-center, as in Figure 5?

Consider this three-part challenge:

(1) Compute the probability that the token will end at the far right if it starts at each of the three interior nodes. (Hint: Set up equations expressing each unknown probability in terms of the other two. Then solve the linear system.)

(2) Use Engel's method to compute these probabilities.

(3) Generalize to a chain with n nodes.

As a final comment, in this article we used the word *node* where Engel used *state*, using terminology from the theory of Markov chains. A full appreciation of Engel's work requires some familiarity with the theory of Markov chains, such as one can obtain from Chapter 11 of Grinstead and Snell's *Introduction to Probability* (Amer. Math. Soc. 2006).

Credit

Chip firing has been rediscovered independently in three different academic communities: mathematics, physics, and computer science. However, it is fitting that its original discovery by Engel was in math education. We often think of mathematical sophistication as something that leads practitioners to create concepts that can be understood only by experts. But at the highest levels of mathematical research, there's a love of clarity that sees the pinnacle of sophistication as the achievement of simplicity where before there was only complexity.

If we want more mathematicians to engage as deeply with pre-college mathematics as Engel did when he taught probability theory at an elementary school or to come up with improved ways of solving problems that we already know how to solve in a more complicated way, we should give the results of those efforts some acclaim. This is all the more true when, as in Engel's case, the fruits of the enterprise proved to be prophetic of later developments in other, seemingly un-related fields of inquiry.

Further Reading

See my blog for an extended discussion of this topic (bit.ly/ProfEngel). On that site, I've posted an authorized copy of Arthur Engel's unpub-lished book *The Stochastic Abacus: An Alternative Approach to Discrete Prob-ability*. See also Engel's two articles—the one mentioned earlier and "The Probabilistic Abacus" (*Educ. Stud. Math.* 6 no. 1 [March 1975]: 1–22).

The On-Line Encyclopedia
of Integer Sequences

Neil J. A. Sloane

Introduction

The *OEIS*® (or *On-Line Encyclopedia of Integer Sequences*®)[1] is a freely accessible database of number sequences, now in its 54th year, and online since 1995. It contains more than 300,000 entries, and for each one gives a definition, properties, references, computer programs, and tables, as appropriate. It is widely referenced: A web page[2] lists more than 6,000 works that cite it; these works often say things like "this theorem would not exist without the help of the *OEIS*." It has been called one of the most useful mathematical sites on the Web.

The main use is to serve as a dictionary or fingerprint file for identifying number sequences (and when you find the sequence you are looking for, you will understand why the *OEIS* is so popular). If your sequence is not recognized, you see a message saying that if the sequence is of general interest, you should submit it for inclusion in the database. The resulting queue of new submissions is a continual source of lovely problems.

I described the *OEIS* in a short article in the September 2003 issue of *Notices of the American Mathematical Society*. The most significant changes since then took place in 2009, when a nonprofit foundation[3] was set up to own and maintain the *OEIS*, and in 2010 when the *OEIS* was moved off my home page at AT&T Labs to a commercial host. The format has also changed: Since 2010, the *OEIS* has been a refereed wiki. Four people played a crucial role in the transition: Harvey P. Dale and Nancy C. Eberhardt helped set up the foundation, Russell S. Cox wrote the software, and David L. Applegate helped move the *OEIS*. The *OEIS* would probably not exist today but for their help.

All submissions of new sequences and updates are now refereed by volunteer editors. One of the rewards of being an editor is that you see a constant flow of new problems, often submitted by nonmathematicians, which frequently contain juicy-looking questions that are begging to be investigated. This article describes a selection of recent sequences, mostly connected with unsolved problems.

Sequences in the *OEIS* are identified by a six-digit number prefixed by A. **A000001** is the number of groups of order n, **A000002** is Kolakoski's sequence, and so on. When we were approaching a quarter of a million entries, the editors voted to decide which sequence would become **A250000**. The winner was the Peaceable Queens sequence, described in the next section, and the runner-up was the "circles in the plane" sequence, **A250001**, discussed after that. The nth term of the sequence under discussion is usually denoted by $a(n)$.

Peaceable Queens

In **A250000**, $a(n)$ is the maximal number m such that it is possible to place m white queens and m black queens on an $n \times n$ chessboard so that no queen attacks a queen of the opposite color. These are peaceable queens. This is a fairly new problem with some striking pictures, an interesting conjecture, and a satisfactorily nonviolent theme. It was posed by Robert A. Bosch in 1999, as a variation on the classical problem of finding the number of ways to place n queens on an $n \times n$ board so that they do not attack each other (**A000170**). It was added to the *OEIS* in 2014 by Donald E. Knuth, and a number of people have contributed to the entry since then. Only 13 terms are known:

n:	1	2	3	4	5	6	7	8	9	10	11	12	13
$a(n)$:	0	0	1	2	4	5	7	9	12	14	17	21	24

Figures 1–4 show examples of solutions for $n = 5, 8, 11$, and (conjecturally) 20.

For larger values of n, the best solutions presently known were found by Benoît Jubin and concentrate the queens into four pentagonal regions, as shown in Figure 5 (and generalize the arrangement shown in Figure 4). This construction gives a lower bound of $\lfloor 7n^2/48 \rfloor$, a formula which in fact matches all the best arrangements known so far

FIGURE 1. One of three solutions to the Peaceable Queens problem on a 5 × 5 board, illustrating $a(5) = 4$. Courtesy of Michael De Vlieger.

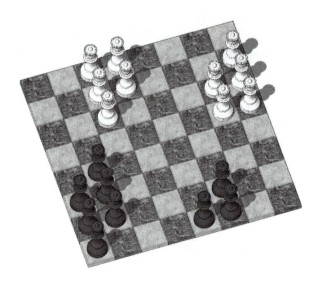

FIGURE 2. A solution to the Peaceable Queens problem on an 8 × 8 board, illustrating $a(8) = 9$. (There are actually 10 white queens here, but only 9 count since the numbers of white and black queens must be equal. Any one of the white queens could be omitted.) Courtesy of Michael De Vlieger.

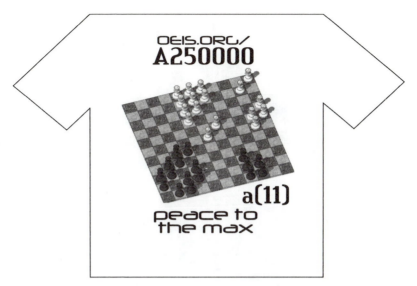

FIGURE 3. A solution to the Peaceable Queens problem on an 11 × 11 board, illustrating $a(11) = 17$. Courtesy of Michael De Vlieger (who also designed the "Peace to the Max" T-shirt).

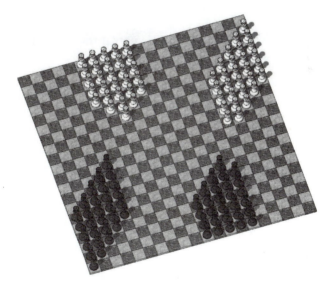

FIGURE 4. A conjectured solution to the Peaceable Queens problem on a 20 × 20 board, found by Bob Selcoe, showing that $a(20) \geq 58$. Courtesy of Michael De Vlieger.

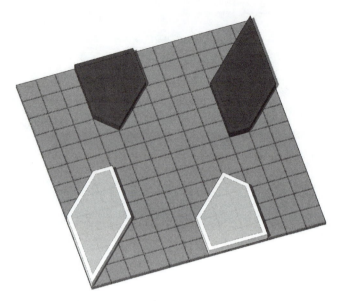

FIGURE 5. A general construction for the Peaceable Queens problem found by Benoît Jubin, showing that for large n, $a(n) \geq \lfloor 7n^2/48 \rfloor$, a formula which might be exact for all $n > 9$. Courtesy of Michael De Vlieger.

except $n = 5$ and 9. It would be nice to know if this construction really does solve the problem!

Circles in the Plane

The runner-up in the competition for **A250000** is now **A250001**: Here $a(n)$ is the number of ways to draw n circles in the affine plane. Two circles must be disjoint or meet in two distinct points (tangential contacts are not permitted), and three circles may not meet at a point.[4] The sequence was proposed by Jonathan Wild, a professor of music at McGill University, who found the values $a(1) = 1$, $a(2) = 3$, $a(3) = 14$, $a(4) = 173$, and, jointly with Christopher Jones, $a(5) = 16951$ (Figures 6–8).

Wild and Jones have found that complications first appear when five circles are being considered: Here there are arrangements that theoretically could exist if one considered only the intersections between circles but that cannot actually be drawn using circles. For example, start with four circles arranged in a chain, each one overlapping its two

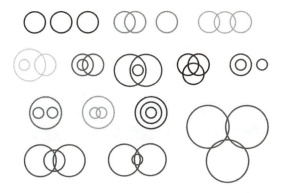

FIGURE 6. The 14 ways to draw three circles in the affine plane. Courtesy of Jessica Gonzalez. See also color images.

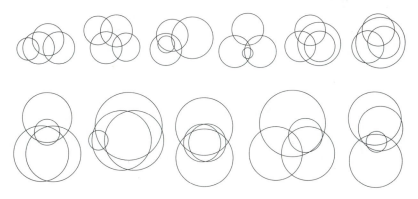

FIGURE 7. Eight of the 173 ways to draw four circles. For the full set of 173 drawings, see **A250001**. Courtesy of Jonathan Wild.

neighbors, and label the overlaps a, b, c, d (Figure 9). Suppose we try to add a fifth circle that meets all four circles but avoids their overlaps, encloses overlaps b and d, but does not enclose overlaps a or c. This figure can be drawn if the fifth circle is flattened to an ellipse, but it can be shown that the arrangement cannot be realized with five circles. There are 26 such unrealizable arrangements of five circles, which can be ruled out by ad hoc arguments.

The delicate configurations like those in Figure 8 are very appealing. It would be interesting to see all 17142 arrangements of five or fewer circles displayed along the Great Wall of China.

FIGURE 8. Seven further ways (out of 173) to draw four circles. Courtesy of Jonathan Wild.

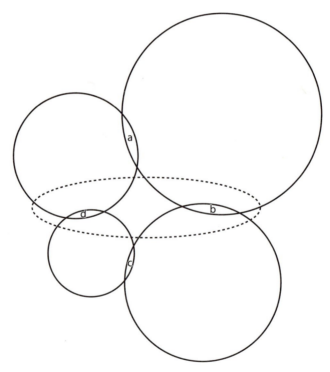

FIGURE 9. A hypothetical arrangement of five circles that can only be realized if one or more of the circles is distorted. Courtesy of Jonathan Wild. See also color images.

Lexicographically Earliest Cube-Free Binary Sequence

There is an obvious way to sort integer sequences $a(1)$, $a(2)$, $a(3)$, $a(4)$, ... into lexicographic order. A number of recent entries in the *OEIS* are defined to be the lexicographically earliest sequence of nonnegative or positive integers satisfying certain conditions.

For example, one of the first results in the subject now called "Combinatorics on Words" was Axel Thue's 1912 theorem that the "Thue-Morse sequence"

$$T = 0, 1, 1, 0, 1, 0, 0, 1, 1, 0, 0, 1,$$
$$0, 1, 1, 0, 1, 0, 0, 1, 0, 1, 1, 0, 0, 1, \ldots$$

(**A010060**) contains no substring of the form XXX, that is, T is *cube-free*. T can be defined as a fixed point of the mapping $0 \to 01$, $1 \to 10$; alternatively, by taking $a(n)$ to be the parity of the number of 1s in the binary expansion of n. One hundred and five years later, David W. Wilson asked for the lexicographically earliest cube-free sequence of 0s and 1s. Using a backtracking algorithm, he found what appear to be the first 10000 terms, which begin

$$0, 0, 1, 0, 0, 1, 0, 1, 0, 0, 1, 0, 0, 1,$$
$$1, 0, 0, 1, 0, 0, 1, 0, 1, 0, 0, 1, \ldots \tag{1}$$

This is now **A282317**.

There is no difficulty in showing that the sequence exists.[5] To see this, make the set S of all infinite binary sequences $a = (a(1), a(2), \ldots)$ into a metric space by defining $d(a, b)$ to be 0 if $a = b$, or 2^{-i} if a and b first differ at position i. This situation identifies S with the Cantor set in $[0, 1)$. The subset $F \subset S$ of infinite cube-free sequences is nonempty and has an infimum c, say. It is easy to show that the complement $S \setminus F$, sequences that contain a cube, is an open set in this topology, so F is closed and $c \in F$.

So far, only the first 999 terms of **A282317** have been verified to be correct (by showing that there is at least one infinite cube-free sequence with that beginning). The rest of the 10000 terms are only conjectural. It would be nice to know more. In particular, does this sequence have an alternative construction? There is no apparent formula or recurrence, which seems surprising.

The EKG and Yellowstone Sequences

To continue the "lexicographically earliest" theme, many recent entries in the *OEIS* are defined to be the lexicographically earliest sequence $a(1)$, $a(2)$, . . . of distinct positive integers satisfying certain divisibility conditions.

The first task here is usually to show that there are no missing numbers, i.e., that the sequence is a permutation of the positive integers. Sequences of this type were studied in a 1983 paper by Erdős, Freud, and Hegyvári [3], which included the examples **A036552** ($a(2n)$ = smallest missing number, $a(2n + 1) = 2a(2n)$) and **A064736** ($a(2n + 2)$ = smallest missing number, $a(2n + 1) = a(2n) \times a(2n + 2)$). For these two, it is clear that there are no missing numbers. This is less obvious, but still true, for Jonathan Ayres's EKG sequence, **A064413**, defined to be the lexicographically earliest sequence of distinct positive integers such that

$$\gcd(a(n - 1), a(n)) > 1 \text{ for all } n \geq 3$$

This sequence begins

$$1, 2, 4, 6, 3, 9, 12, 8, 10, 5, 15, 18,$$
$$14, 7, 21, 24, 16, 20, 22, 11, 33, 27, \ldots$$

The proof that it is a permutation is omitted—it is similar to the proof for the Yellowstone sequence given below.

Next, one can investigate the rate of growth. In the case of **A064413**, the points appear to lie roughly on three curved lines (Figure 10), although the following conjecture of Lagarias, Rains, and Sloane [6] is still open.

CONJECTURE 1. *In the EKG sequence* **A064413**, *if $a(n)$ is neither a prime nor three times a prime, then*

$$a(n) \sim n\left(1 + \frac{1}{3 \log n}\right);$$

if $a(n)$ is a prime, then

$$a(n) \sim \frac{1}{2}n\left(1 + \frac{1}{3 \log n}\right);$$

and if $a(n)$ is 3 times a prime, then

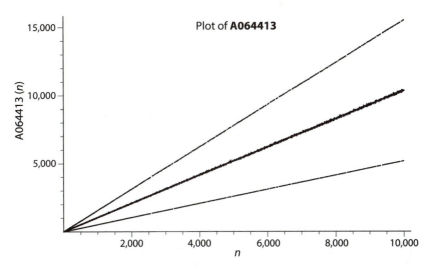

FIGURE 10. The first 10000 terms of the EKG sequence, so named because locally this graph resembles an EKG. Every number appears exactly once. Courtesy of Michael De Vlieger. See also color images.

$$a(n) \sim \frac{3}{2}n\left(1 + \frac{1}{3\log n}\right).$$

Furthermore, if the sequence is a permutation, one can also try to study its cycle structure. However, this sequence often leads to difficult questions, similar to those encountered in studying the Collatz conjecture, and we can't do much more than collect experimental data. Typically, there is a set of finite cycles, and one or more apparently infinite cycles, but we can't prove that the apparently infinite cycles really are infinite, nor that they are distinct. See the entries for **A064413** and **A098550** for examples.

The definition of the Yellowstone sequence (Reinhard Zumkeller 2004, **A098550**, [1]) is similar to that of the EKG sequence, but now the requirement is that, for $n > 3$,

$$\gcd(a(n - 2), a(n)) > 1 \text{ and } \gcd(a(n - 1), a(n)) = 1.$$

This sequence begins

$$1, 2, 3, 4, 9, 8, 15, 14, 5, 6, 25, 12, 35,$$
$$16, 7, 10, 21, 20, 27, 22, 39, 11, \ldots$$

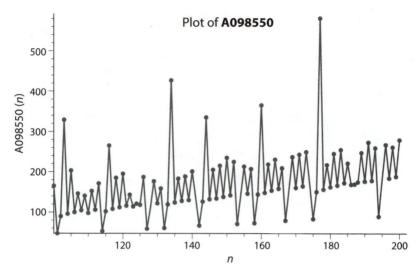

FIGURE 11. Plot of terms $a(101)$ through $a(200)$ of the Yellowstone sequence. The sequence has a downward spike to $a(n)$ when $a(n)$ is a prime and larger upward spikes (the "geysers," which suggests the name for this sequence) two steps later. Courtesy of Michael De Vlieger. See also color images.

Figure 11 shows terms $a(101) = 47$ through $a(200) = 279$, with successive points joined by lines.

THEOREM 2. *The Yellowstone sequence* **A098550** *is a permutation of the positive integers.*

The proof is typical of the arguments used to prove that several similar sequences are permutations, including the EKG sequence above.

Proof. *There are several steps.*

(i)   The sequence is infinite. (For $pa(n - 2)$ is always a candidate for $a(n)$, where p is a prime larger than any divisor of $a(i)$, $i < n$.)

(ii) There are infinitely many different primes that divide the terms of the sequence. (If not, there is a prime p such that all terms are products of primes less than p. Using (i), find a term $a(n) > p^2$, and let q be a common prime factor of $a(n - 2)$ and $a(n)$. But now $pq < p^2 < a(n)$ is a smaller candidate for $a(n)$, a contradiction.)

(iii) For any prime p, some term is divisible by p. (If not, no prime $q > p$ can divide any $a(n)$: If $a(n) = kq$ is the first multiple of q to appear, kp would be a smaller candidate for $a(n)$. This contradicts (ii).)

(iv) For any prime p, p divides infinitely many terms. (If not, let p^i be larger than any multiple of p in the sequence, and choose a prime $q > p^i$. Again we obtain a contradiction.)

(v) Every prime p is a term in the sequence. (Suppose not, and using (i), choose n_0 such that $a(n) > p$ for all $n > n_0$. Using (iv), find $a(n) = kp$, $k > 1$, for some $n > n_0$. But then $a(n + 2) = p$, a contradiction.)

(vi) All numbers appear. If not, let k be the smallest missing number, and choose n_0 so that all of $1, \ldots, k - 1$ have occurred in $a(1), \ldots, a(n_0)$. Let p be a prime dividing k. Since, by (iv), p divides infinitely many terms, there is a number $n_1 > n_0$ such that $\gcd(a(n_1), k) > 1$. This forces

$$\gcd(a(n), k) > 1 \text{ for } \textbf{all } n \geq n_1 \qquad (2)$$

(If not, there would be some $j \geq n_1$ where $\gcd(a(j), k) > 1$ and $\gcd(a(j + 1), k) = 1$, which would lead to $a(j + 2) = k$.) But (2) is impossible, because we know from (v) that infinitely many of the $a(n)$ are primes. $\square$

The growth of this sequence is more complicated than that of the EKG sequence. Figure 12 shows the first 300,000 terms, without lines connecting the points. The points appear to fall on or close to a number of distinct curves. There is a conjecture in [1, p. 5] that would explain these curves.

Three Further Lexicographically Earliest Sequences

Here are three further examples of this type, all of which are surely permutations of the positive integers. For the first there is a proof, for the second there is "almost" a proof, but the third may be beyond reach.

The first example (Leroy Quet 2007, **A127202**) is the lexicographically earliest sequence of distinct positive integers such that

$$\gcd(a(n - 1), a(n)) \neq \gcd(a(n - 2), a(n - 1)) \text{ for } n \geq 3$$

It begins

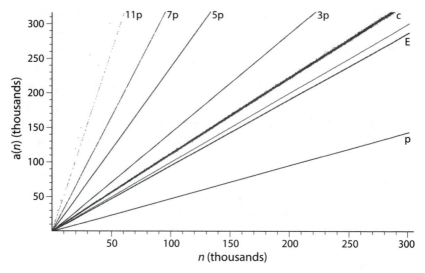

FIGURE 12. Scatter plot of the first 300,000 terms of the Yellowstone sequence. The primes lie on the lowest line (labeled "p"), the even numbers on the second line ("E"), the majority of the odd composite numbers on the third line ("C"), and the $3p$, $5p$, $7p$, $11p$, ... points on the higher lines. The lines are not actually straight, except for the line $f(x) = x$ (located between the line labeled "c" and the line labeled "E"), which is included for reference. Courtesy of Hans Havermann. See also color images.

$$1, 2, 4, 3, 6, 5, 10, 7, 14, 8, 9, 12, 11, 22,$$
$$13, 26, 15, 18, 16, 17, 34, 19, \ldots$$

For the second example (Rémy Sigrist 2017, **A280864**), the definition is the following:

> If a prime p divides $a(n)$, then it divides exactly
> one of $a(n - 1)$ and $a(n + 1)$, for $n \geq 2$

and the initial terms are

$$1, 2, 4, 3, 6, 8, 5, 10, 12, 9, 7, 14, 16, 11,$$
$$22, 18, 15, 20, 24, 21, 28, 26, \ldots$$

The proof that the first is a permutation is similar to that for the Yellowstone sequence, although a bit more involved (see **A127202**). The second example struck me as one of those "drop everything and work on this" problems that are common hazards when editing new submissions to the *OEIS*. However, after several months, I could prove that every prime and every even number appears, and that if p is an odd prime, then there are infinitely many odd multiples of p (see **A280864** for details), but I could not prove that every odd number appears. The missing step feels like it is only a couple of cups of coffee away, and I'm hoping that some reader of this article will complete the proof.

The third example (Henry Bottomley 2000, **A055265**) is the lexicographically earliest sequence of distinct positive integers such that $a(n-1) + a(n)$ is a prime for $n \geq 2$:

$$1, 2, 3, 4, 7, 6, 5, 8, 9, 10, 13, 16, 15,$$
$$14, 17, 12, 11, 18, 19, 22, 21, 20, \ldots$$

The terms appear to lie on or near the line $a(n) = n$, but the proof that every number appears may be difficult because it involves the gaps between the primes.

Two-Dimensional Lexicographically Earliest Arrays

The *OEIS* is primarily a database of sequences $(a_n, n \geq n_0)$. However, triangles of numbers are included by reading them by rows. Pascal's triangle becomes 1, 1, 1, 1, 2, 1, 1, 3, 3, 1, 1, 4, 6, 4, 1, . . . , which (without the extra spaces) is **A007318**. Doubly indexed arrays $(T_{m,n}, m \geq m_0, n \geq n_0)$ are converted to sequences by reading them by antidiagonals (in either the upward or downward direction, or both). So an array $(T_{m,n}, m \geq 0, n \geq 0)$ might become $T_{0,0}, T_{1,0}, T_{0,1}, T_{2,0}, T_{1,1}, T_{0,2}, \ldots$. For example, the table of Nim sums $m \oplus n$:

0	1	2	3	4	5	6	7	. . .
1	0	3	2	5	4	7	6	. . .
2	3	0	1	6	7	4	5	. . .
3	2	1	0	7	6	5	4	. . .
4	5	6	7	0	1	2	3	. . .
5	4	7	6	1	0	3	2	. . .
6	7	4	5	2	3	0	1	. . .
7	6	5	4	3	2	1	0	. . .
•	•	•	•	•	•	•	•	. . .

$$(3)$$

produces the sequence **A003987**.

$$0, 1, 1, 2, 0, 2, 3, 3, 3, 3, 4, 2, 0, 2, 4,$$
$$5, 5, 1, 1, 5, 5, 6, 4, 6, 0, 6, 4, 6, \ldots$$

Doubly indexed, doubly infinite arrays ($T_{m,n}$, $m \in \mathbb{Z}$, $n \in \mathbb{Z}$) can become sequences by reading them in a spiral around the origin, in say a counterclockwise direction: $T_{0,0}$, $T_{1,0}$, $T_{1,1}$, $T_{0,1}$, $T_{-1,1}$, $T_{-1,0}$, $T_{-1,-1}$, $T_{0,-1}$, $\ldots$ (Figure 13).

There are many "lexicographically earliest" versions of these arrays. For example, the Nim sum array (3) has an equivalent definition: scan along upward antidiagonals, filling in each cell with the smallest nonnegative number that is neither in the row to the left of that cell nor in the column above it.

```
9  ← 16 ←  2 ←  4 ←  7 ← 14 ← 11 ← 12 ←  1 ←  5 ←  8
↓                                                    ↑
17    8 ← 15 ← 14 ← 13 ← 12 ←  9 ← 10 ←  6 ←  7     3
↓    ↓                                    ↑     ↑
1    2    4 ← 11 ← 10 ←  3 ←  8 ←  7 ←  9    13    15
↓    ↓    ↓                          ↑     ↑     ↑
8    9    7    3 ←  5 ←  6 ←  1 ←  2     4    12    11
↓    ↓    ↓    ↓                    ↑    ↑    ↑     ↑
11   12    8    1    2 ←  4 ←  3     6     5    10    14
↓    ↓    ↓    ↓    ↓         ↑    ↑    ↑    ↑     ↑
15    7    6    5    3    1 →  2     4     8    11    12
↓    ↓    ↓    ↓    ↓              ↑    ↑    ↑     ↑
14   10    3    2    4 →  5 →  6 →  1     7     9    13
↓    ↓    ↓    ↓                    ↑    ↑     ↑
7    11    9    6 →  1 →  2 →  4 →  5 →  3     8    10
↓    ↓    ↓                                ↑     ↑
4    13    5 →  7 →  8 →  9 → 10 → 11 → 12 →  6     1
↓    ↓                                          ↑
12   14 → 10 →  9 →  6 → 13 →  5 →  3 → 15 → 16 →  7
↓
10 → 15 →  1 → 12 → 16 →  8 → 14 → 13 → 11 → 18 → 17
```

FIGURE 13. **A274640**: Choose the smallest positive number so that no row, column, or diagonal contains a repeat. Are the rows, columns, and diagonals permutations of $\mathbb{N}$? Courtesy of Michael De Vlieger.

A variation on the Nim sum array was proposed by Alec Jones in 2016, as a kind of "infinite Sudoku array." This array ($T_{m,n}$, $m \geq 0$, $n \geq 0$) is to be filled in by upward antidiagonals, always choosing the smallest positive integer such that no row, column, diagonal, or antidiagonal contains a repeated term. The top left corner of the array is

$$
\begin{array}{cccccccccc}
1 & 3 & 2 & 6 & 4 & 5 & 10 & 11 & \ldots \\
2 & 4 & 5 & 1 & 8 & 3 & 6 & 12 & \ldots \\
3 & 1 & 6 & 2 & 9 & 7 & 5 & 4 & \ldots \\
4 & 2 & 3 & 5 & 1 & 8 & 9 & 7 & \ldots \\
5 & 7 & 1 & 4 & 2 & 6 & 3 & 15 & \ldots \\
6 & 8 & 9 & 7 & 5 & 10 & 4 & 16 & \ldots \\
7 & 5 & 4 & 3 & 6 & 14 & 8 & 9 & \ldots \\
8 & 6 & 7 & 9 & 11 & 4 & 13 & 3 & \ldots \\
\bullet & \bullet & \bullet & \bullet & \bullet & \bullet & \bullet & \bullet & \ldots
\end{array}
\tag{4}
$$

The resulting sequence (**A269526**) is

$$1, 2, 3, 3, 4, 2, 4, 1, 5, 6, 5, 2, 6, 1, 4,$$
$$6, 7, 3, 2, 8, 5, 7, 8, 1, 5, 9, 3, 10, \ldots$$

This array has many interesting properties. If we subtract 1 from each entry, the entries are the Nim values for a game played with two piles of counters, of sizes m and n, and reminiscent of Wythoff's game (see **A004481**, **A274528**).

But the main question about the array (4) is, are the individual rows, columns, and diagonals of this array permutations of $\mathbb{N}$? (The antidiagonals are obviously not, since they are finite sequences.) It is easy to see that each column is a permutation. In column $c \geq 0$, a number k will eventually be the smallest missing number and will appear in some cell in that column, unless there is a copy of k to the northwest, west, or southwest of that cell. But there are at most c copies of k in all the earlier columns, so eventually k *will* appear.

The rows are also permutations, although the proof is less obvious. Consider row $r \geq 0$, and suppose that k never appears. There are at most r copies of k in the earlier rows, and these can affect only a bounded portion of row r. Consider a cell (r,n), $n \geq 0$ large. If k is not to appear in that cell, there must be a copy of k in the antidiagonal to the southwest. So in the triangle bounded by row r, column 0, and the antidiagonal through (r,n), there must be at least $n + 1 - r$ copies of k. Imagine these *ks* replaced by chess queens. By construction, they are mutually nonattacking. But it is known ([9, Problem 252], or **A274616**) that on a triangular half-chessboard of side n, there can be at most $2n/3 + 1$ mutually nonattacking queens, which for large n leads to a contradiction.

As to the diagonals, although they appear to be permutations, this is an open question. The argument using nonattacking queens breaks down because the diagonal of the half-chessboard contains only half as many squares as the sides. Even the main diagonal, **A274318**,

$$1, 4, 6, 5, 2, 10, 8, 3, 7, 9, 16, 26, 29,$$
$$22, 20, 23, 28, 38, 12, 32, 46, 13, 14, 11, 15, \ldots$$

is not presently known to be a permutation of N.

The spiral version of this array is even more frustrating. This array $((T(m,n), m \in \mathbb{Z}, n \in \mathbb{Z})$, **A274640**, proposed by Zak Seidov and Kerry Mitchell in June 2016), is constructed in a counterclockwise spiral, filling in each cell with the smallest positive number such that no row, column, or diagonal contains a repeated term (Figures 13, 14). ("Diagonal" now means any line of cells of slope ±1.)

Although it seems very plausible that every row, column, and diagonal is a permutation of N, now there are no proofs at all. The eight spokes through the center are sequences **A274924–A274931**. For example, the row through the central cell is

$$\ldots, 14, 25, 13, 17, 10, 15, 7, 6, 5, 3,$$
$$1, 2, 4, 8, 11, 12, 16, 9, 19, 24, 22, \ldots,$$

which is **A274928** reversed followed by **A274924**. Is it a permutation of N? We do not know.

Fun with Digits

Functions of the digits of numbers have always fascinated people,[6] and one such function was in the news in 2017. The idea underlying this story and several related sequences is to start with some simple function $f(n)$ of the digits of n in some base, iterate it, and watch what happens.

For the first example, we write n as a product of prime powers, $n = \ldots$ with the p_i in increasing order, and define $f(n)$ to be the decimal concatenation $p_1^{e_1} p_2^{e_2} \ldots$, where we omit any exponents e_i that are equal to 1. So $f(7) = f(7^1) = 7, f(8) = f(2^3) = 23$.

FIGURE 14. Colored representation of central 200×200 portion of the spiral in Figure 13: the colors represent the values, ranging from black (smallest) to white (largest). Courtesy of Kerry Mitchell. See also color images.

The initial values of $f(n)$ (**A080670**) are

n:	1	2	3	4	5	6	7	8	9	10	11	12	13	14	15	16	17	18	...
$f(n)$:	1	2	3	22	5	23	7	23	32	25	11	223	13	27	35	24	17	232	...

If we start with a positive number n and repeatedly apply f, in many small cases we rapidly reach a prime (or 1).[7] For example, $9 = 3^2 \to 32 = 2^5 \to 25 = 5^2 \to 52 = 2^2 13 \to 2213$, a prime. Define $F(n)$ to be the prime that is eventually reached, or -1 if the iteration never reaches a prime. The value -1 will occur if the iterates are unbounded or if they enter a cycle of composite numbers. The initial values of $F(n)$ (**A195264**) are

$$1, 2, 3, 211, 5, 23, 7, 23, 2213, 2213, 11,$$
$$223, 13, 311, 1129, 233, 17, 17137, 19, \ldots .$$

$F(20)$ is currently unknown (after 110 steps, the trajectory of 20 has stalled at a 192-digit number that has not yet been factored). At a conference at the Center for Discrete Mathematics and Theoretical Computer Science (DIMACS) in October 2014, to celebrate the 50th anniversary of the start of what is now the *OEIS*, John H. Conway offered $1,000 for a proof or disproof of his conjecture that the iteration of f will always reach a prime.

However, in June 2017, James Davis found a number $D_0 = 13532385396179$ whose prime factorization is $13 \times 53^2 \times 3853 \times 96179$, and so clearly $f(D_0) = D_0$ and $F(D_0) = -1$.

The method used by James Davis to find D_0 is quite simple. Suppose $n = m \times p$ is fixed by f, where p is a prime greater than all the prime factors of m. Then $f(n) = f(m) 10^y + p$, where y is the number of digits in p. From $f(n) = n$ we have $p = \frac{f(m)10^y}{m-1}$. Assuming $p \neq 2, 5$, this implies that p divides $f(m)$, and setting $x = f(m)/p$, we find that $m = x10^y + 1$ with $p = \frac{f(m)}{x}$ prime. A computer easily finds the solution $x = 1407$, $y = 5$, $m = 140700001$, $p = 96179$, and so $n = D_0$.

No other composite fixed points are known, and David J. Seal (private communication) has recently shown that there is no composite fixed point less than D_0. It is easy, however, to find numbers whose trajectory under f ends at D_0, by repeatedly finding a prime prefix of the previous number, as shown by the example[8] $D_1 = 13^{532385396179}$ with $f(D_1) = D_0$. So presumably, there are infinitely many n with $F(n) = -1$.

Consideration of the analogous questions in other bases might have suggested that counterexamples to Conway's question could exist. We use subscripts to indicate the base (so $4_{10} = 100_2$). The base-2 analog of f, f_2 (say), is defined by taking $f_2(p_1^{e_1} p_2^{e_2} \ldots)$ to be the concatenation $p_1 e_1 p_2 e_2 \ldots$, as before (again omitting any e_i that are 1), except that now we write the p_i and e_i in base 2 and interpret the concatenation as a base-2 number. For example, $f_2(8) = f_2(2^3) = 1011_2 = 11_{10}$.

The initial values of $f_2(n)$ (**A230625**) are

n:	1	2	3	4	5	6	7	8	9	10	11	12	13	14	15	16	17	18 $\ldots$
$f_2(n)$:	1	2	3	10	5	11	7	11	14	21	11	43	13	23	29	20	17	46 $\ldots$

and the base-2 analog of F, F_2 (**A230627**) is the prime (or 1) that is reached when f_2 is repeatedly applied to n, or -1 if no prime (or 1) is reached:

1, 2, 3, 31, 5, 11, 7, 11, 23, 31, 11, 43, 13, 23, 29, 251, 17, 23, . . .

Now there is a fairly small composite fixed point, namely 255987, found by David J. Seal. Sean A. Irvine and Chai Wah Wu have also studied this sequence, and the present status is that $F_2(n)$ is known for all n less than 12388. All numbers in this range reach 1, a prime, the composite number 255987, or one of the two cycles 1007 ↔ 1269 or 1503 ↔ 3751. The numbers for which $F_2(n) = -1$ are 217, 255, 446, 558, . . . (**A288847**). Initially it appeared that 234 might be on this list, but Irvine found that after 104 steps the trajectory reaches the 51-digit prime

350743229748317519260857777660944018966290406786641

Home Primes

A rather older problem arises if we change the definition of $f(n)$ slightly, making $f(8) = 222$ rather than 23. So if $n = p_1 \times p_2 \times p_3 \times \ldots$, where $p_1 \leq p_2 \leq p_3 \leq \ldots$, then $f(n)$ is the decimal concatenation $p_1 p_2 p_3 \cdots$ (**A037276**). In 1990, Jeffrey Heleen studied the analog of $F(n)$ for this function: that is, $F(n)$ is the prime reached if we start with (n) and repeatedly apply f, or -1 if no prime is ever reached (**A037274**).

The trajectory of 8 now takes 14 steps to reach a prime (the individual prime factors here have been separated by spaces):

$$8 \rightarrow 2\ 2\ 2 \rightarrow 2\ 3\ 37 \rightarrow 3\ 19\ 41 \rightarrow 3\ 3\ 3\ 7\ 13\ 13 \rightarrow 3\ 11123771$$
$$\rightarrow 7\ 149\ 317\ 941 \rightarrow\rightarrow 229\ 31219729 \rightarrow 11\ 2084656339$$
$$\rightarrow 3\ 347\ 911\ 118189 \rightarrow 11\ 613\ 496501723 \rightarrow$$
$$\rightarrow 97\ 130517\ 917327 \rightarrow 53\ 1832651281459$$
$$\rightarrow 3\ 3\ 3\ 11\ 139\ 653\ 3863\ 5107$$
$$\rightarrow 3331113965338635107$$

The last number is a prime.

Since $f(n) > n$ if n is composite, now there cannot be any composite fixed points nor any cycles of length greater than 1. The only way for $F(n)$ to be -1 is for the trajectory of n to be unbounded. This appears to be a harder problem than the one in the previous section, since so far no trajectory has been proved to be unbounded. The first open case is $n = 49$, which after 119 iterations has reached a 251-digit composite number (see **A056938**). The completion of the factorization for step 117 took 765 days by the general number field sieve, and at the time (December 2014) was one of the hardest factorizations ever completed.

Power Trains

A third choice for $f(n)$ was proposed by John H. Conway in 2007: he called it the *power train* map. If the decimal expansion of n is $d_1 d_2 d_3 \ldots d_k$ (with $0 \le d_i \le 9$, $0 < d_1$), then $f(n) = d_1^{d_2} \times d_3^{d_4} \ldots$, ending with $\ldots d_k$ if k is odd, or with $\ldots d_{k-1}^{d_k}$ if k is even (**A133500**). We take 0^0 to be 1. For example, $f(39) = 3^9 = 19683$, $f(623) = 6^2 \times 3 = 108$. Conway observed that $2592 = 2^5 9^2$ is a nontrivial fixed point, and asked me if there were any others. I found one more: $n = 2^{46} \times 3^6 \times 5^{10} \times 7^2 = 24547284284866560000000000$, for which $f(n) = 2^4 \times 5^4 \times 7^2 \times 8^4 \times 2^8 \times 4^8 \times 6^6 \times 5^6 \times 0^0 \times 0^0 \times 0^0 \times 0^0 \times 0^0 = n$. The eleven known fixed points (including the trivial values $1, \ldots, 9$) form **A135385**, and it is known that there are no further terms below 10^{100}. Maybe this is a hint that for all of the functions $f(n)$ that have just been mentioned, there may be only a handful of genuinely exceptional values?

A Memorable Prime

If you happen to need an explicit 20-digit prime in a hurry, it is useful to remember that although 1, $121 = 11^2$, $12321 = 111^2$, $1234321 = 1111^2$, $\ldots$, and $12345678987654321 = 111111111^2$ are not primes, the next term in **A173426** *is* a prime,

$$12345678910987654321$$

As David Broadhurst remarked on the Number Theory Mailing List in August 2015, this is a memorable prime! He also pointed out that on probabilistic grounds, there should be infinitely many values of n such that the decimal concatenation of the numbers 1 up through n followed by $n - 1$ down through 1 is a prime. Shortly afterward, Shyam Sunder Gupta found what is presumably the next prime in the sequence, corresponding to $n = 2446$, the 17350-digit probable prime 1234567. .2445 24462445. .7654321. Serge Batalov has shown that there are no further terms with $n < 60000$. What is the next term? The values 10 and 2446 are not enough to create an *OEIS* entry.

A Missing Prime

The previous question naturally led me to wonder what the first prime is in the simpler sequence (**A007908**):

$$1, 12, 123, 1234, \ldots, 12345678910, 1234567891011, \ldots$$

formed by the decimal concatenation of the numbers 1 through n. In *Unsolved Problems in Number Theory* [5], Richard K. Guy reports that this question was already asked by Charles Nicol and John Selfridge. However, although the same probabilistic argument suggests that there should be an infinite number of primes of this type, not a single one is known. I asked several friends to help with the search, and as a result this task was taken up by the folks who run the GIMP (or Great Internet Mersenne Prime) search, and there is now a web page[9] that shows the current status of the search for the first prime. As of August 2017, the search seems to have stalled, the present status being that all the potential values of n through 344869 failed (obviously many values of n can be ruled out by congruence conditions). In this range. the candidates have about 2 million digits. One estimate suggests that there is a probability of about 0.5 that a prime will be found with $n < 10^6$, so it would be good to resume this search.

Post's Tag System

In his recent book *Elements of Mathematics: From Euclid to Gödel*[10] John Stillwell [8] mentions that Emil L. Post's tag system from the 1930s is still not understood. Post asked the following question. Take a finite string, or word, S of 0s and 1s, and if it begins with 0, append 00 to the end of S and delete the first three symbols, or if it begins with 1, append 1101 to the end of S and delete the first three symbols. When this process is iterated, eventually one of three things will happen: either S will reach the empty word (S *dies*), S will enter a loop (S *cycles*), or S will keep growing forever (S *blows up*). For example, $S = 1000$ reaches the empty word ε at the seventh step:

$$1000 \rightarrow 01101 \rightarrow 0100 \rightarrow 000 \rightarrow 00 \rightarrow 0 \rightarrow \varepsilon$$

whereas 100100 enters a cycle of length six (indicated by parentheses) after 15 steps:

$$
\begin{aligned}
100100 &\rightarrow 1001101 \rightarrow 11011101 \rightarrow 111011101 \\
&\rightarrow 0111011101 \rightarrow 101110100 \rightarrow 1101001101 \\
&\rightarrow 10011011101 \rightarrow 110111011101 \\
&\rightarrow 1110111011101 \rightarrow 01110111011101 \\
&\rightarrow 1011101110100 \rightarrow 11011101001101
\end{aligned}
\tag{5}
$$

$$\to 111010011011101 \to 0100110111011101$$
$$\to (011011101110100 \to 01110111010000$$
$$\to 1011101000000 \to 11010000001101$$
$$\to 100000011011101 \to 0000110111011101)$$

Post was hoping to find an algorithm which, given S, would determine which of these outcomes would occur. He did not succeed.

Post called this process a "tag system." It can be generalized by considering initial words over an alphabet of size M (rather than 2), allowing any fixed set $\mathcal{A}$ of M tag words to be appended (rather than 00 and 1101), and deleting some fixed number P of initial symbols at each step (not necessarily 3). In 1961, Marvin Minsky showed that such a generalized tag system could simulate a Turing machine. By choosing an appropriate alphabet, an appropriate set $\mathcal{A}$ of tag words to be appended, and an appropriate value of P (in fact $P = 2$ will do), any computable function can be simulated. So, because of the undecidability of the Halting Problem, for general tag systems it is impossible to predict which initial words will blow up.

But what about Post's original tag system? Could this simulate a Turing machine (by encoding the problem in the initial word S)? At first this seems very unlikely, but the Cook–Wolfram theorem that the one-dimensional cellular automaton defined by Rule 110 can simulate a Turing machine (by encoding the problem in the starting state) suggests that it might be possible. If it *is* possible, there must be some initial words that blow up (again because of the Halting Problem).

In early 2017, when I read Stillwell's book, the *OEIS* contained three sequences related to the original tag system, based on the work of Peter Asveld and submitted by Jeffrey Shallit: **A284116**, giving the maximal number of words in the "trajectory" of any initial word S of length n (18 terms were known), and two sequences connected with the especially interesting starting word σ_n of length $3n$ consisting of n copies of 100. **A284119**(n) is defined to be the number of words in the trajectory of σ_n before it enters a cycle or dies, or -1 if the trajectory blows up, and **A284121**(n) is the length of the cycle, or 1 if the trajectory dies, or -1 if the trajectory blows up. For example, from (5) we see that **A284119**$(2) = 15$ and **A284121**$(2) = 6$. Shallit had extended Asveld's work and had found 43 terms of the two last-mentioned sequences.

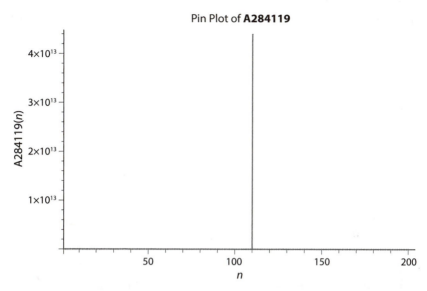

Pin Plot of **A284119**

FIGURE 15. Pin plot illustrating Lars Blomberg's remarkable discovery that the Post tag system started at the word $(100)^{110}$ takes an exceptionally long time (43913328040672 steps) to converge. Courtesy of Michael De Vlieger.

I then added many further sequences based on tag systems discussed by Asveld, Liesbeth De Mol, Shigeru Watanabe, and others, and appealed to contributors to the *OEIS* to extend them.

The most interesting response came from Lars Blomberg, who investigated the trajectory of σ_n for $n \leq 110$. On September 9, 2017, he reported that every σ_n for $n \leq 110$ had either died or cycled after at most 13 million terms, except for σ_{110}, which after 38.10^{11} steps had reached a word of length 10^7 and was still growing. This was exciting news! Could σ_{110} be the first word to be discovered that blew up?[11] Sadly, on October 4, 2017, Blomberg reported that after 43913328040672 steps σ_{110} had terminated in the empty word.

Figure 15 displays the remarkable graph (technically, a pin plot) of the number of steps for σ_n to either die or cycle for $n \leq 200$. Figure 16 shows the lengths of the successive words in the trajectory of σ_{110}.

In the past six months, Blomberg has continued this investigation and has determined the fate of σ_n for all $n \leq 6075$. The new record holder for the number of steps before the trajectory dies is now held by σ_{4974}, which takes 570422251906801 steps, while σ_{110} is in second place.

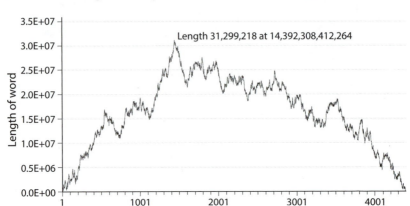

FIGURE 16. Lengths of successive words in trajectory of $(100)^{110}$ under the Post tag system. The numbers on the horizontal axis are spaced at multiples of 10^{12}. Courtesy of Lars Blomberg.

Of course, it is still possible that some initial word S, not necessarily of the form σ_n, will blow up, but this possibility seems increasingly unlikely. So Post's tag system probably does not simulate a Turing machine.

The question as to which σ_n die and which cycle remains a mystery. Up to $n = 6075$, Blomberg's results show that about one-sixth of the values of n die and five-sixths cycle. The precise values can be found in **A291792**. It would be nice to understand this sequence better.

Coordination Sequences

This final section is concerned with *coordination sequences*, which arise in crystallography and in studying tiling problems; they have beautiful illustrations and lead to many unsolved mathematical questions.

The "Cairo" tiling, so called because it is said to be used on many streets in that city, is shown in Figure 17. Let G denote the corresponding infinite graph (with vertices for points where three or more tiles meet, and edges between two vertices where two tiles meet). The figure is also a picture of the graph.

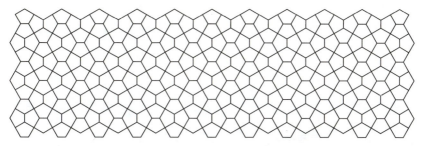

FIGURE 17. A portion of the Cairo tiling. Courtesy of Chaim Goodman-Strauss.

The distance between vertices P, $Q \in G$ is defined to be the number of edges in the shortest path joining them. The *coordination sequence* of G with respect to a vertex $P \in G$ is then the sequence $a(n)$ $(n \geq 0)$ giving the number of vertices Q at distance n from P. Coordination sequences have been studied by crystallographers for many years [7].

The graph of the Cairo tiling has two kinds of vertices, trivalent (where three edges meet) and tetravalent. As can be seen from Figure 17, the coordination sequence with respect to a tetravalent vertex begins 1, 4, 8, 12, 16, 20, 24, . . . , which appears to be the same as the coordination sequence **A008574** for a vertex in the familiar square grid. This observation seemed to be new. Chaim Goodman-Strauss and I thought that such a simple fact should have a simple proof, and we developed an elementary "coloring book" procedure [4], which not only proved this result but also established a number of conjectured formulas for coordination sequences of other tilings mentioned in entries in the *OEIS*. The "coloring book" drawing of the Cairo graph centered at a tetravalent vertex is shown in Figure 18. This coloring makes it easy to prove that the coordination sequence is given by $a(n) = 4n$ for $n \geq 1$ (see [4] for details).

For a trivalent vertex in the Cairo tiling, the coordination sequence is

1, 3, 8, 12, 15, 20, 25, 28, 31, 36, 41, 44, 47, 52, 57, 60, 63, 68, . . .

(this is now **A296368**), and we [4] show that for $n \geq 3$, $a(n) = 4n$ if n is odd, $4n - 1$ if $n \equiv 0 \pmod 4$, and $4n + 1$ if $n \equiv 2 \pmod 4$.

One can similarly define coordination sequences for other two- and higher-dimensional structures, and the *OEIS* currently contains more than 7000 such sequences (mostly without formulas). Many more could be added. There are many excellent websites with lists of tilings and

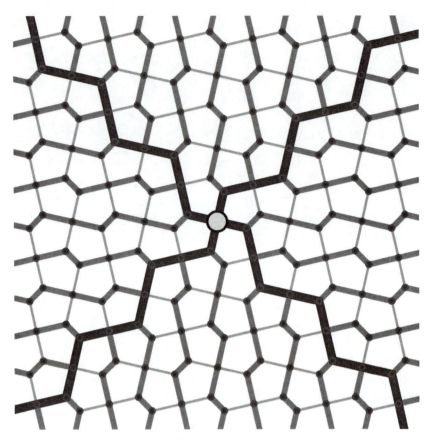

FIGURE 18. The "coloring book" method applied to a tetravalent vertex (the large dot in the center) in the Cairo tiling, used to prove that the coordination sequence is the same as that for the square grid. Courtesy of Chaim Goodman-Strauss. See also color images.

crystals. Brian Galebach's website[12] is especially important, as it includes pictures of all "k-uniform" tilings with $k \leq 6$, with more than 1000 tilings. Darrah Chavey's article [2] and the Michael Hartley and Printable Paper websites[13] have many further pictures, and the RCSR and ToposPro databases[14] have thousands more.

Only last week (on May 4, 2018), Rémy Sigrist investigated the Ammann–Beenker (or "octagonal") tiling shown in Figure 19, an aperiodic tiling with eight-fold rotational symmetry about the central point.

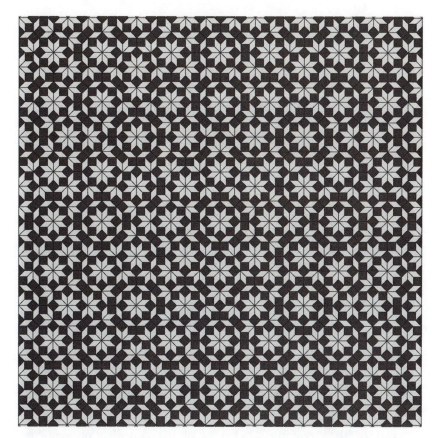

FIGURE 19. Ammann-Beenker (or "octagonal") tiling from *The Tilings Encyclopedia*, https://tilings.math.uni-bielefeld.de. Used under the Creative Commons Attribution-NonCommercial-ShareAlike 2.0 Generic License. See also color images.

Sigrist determined the initial terms of the coordination sequence with respect to the central vertex (**A303981**):

$$1, 8, 16, 32, 32, 40, 48, 72, 64, 96, 80, 104, 112, 112, 128, 152, \ldots \tag{6}$$

Figure 20 shows the vertexes at distances 0, 1, 2, . . . , 6 from the center.

No formula or growth estimate is currently known for this sequence. However, earlier this year, Anton Shutov and Andrey Maleev

FIGURE 20. Illustrating the coordination sequence for the Ammann-Beenker tiling, showing the vertices at distances 0 though 6 from the central vertex. Courtesy of Rémy Sigrist. See also color images.

determined the asymptotic behavior of the coordination sequence (**A302176**) with respect to a vertex with fivefold rotational symmetry in a certain Penrose tiling. So we end with a question: Can the Shutov-Maleev approach be used to find the asymptotic growth of (6)? Of course, an explicit formula would be even nicer.

Acknowledgments

Now that the *OEIS* is a wiki, many volunteer editors help maintain it.[15] Besides the *OEIS* contributors already mentioned in this article, I would like to thank Jörg Arndt, Michael De Vlieger, Charles Greathouse IV, Alois P. Heinz, Michel Marcus, Richard J. Mathar, Don Reble, Jon E. Schoenfield, and Allan C. Wechsler for their help. (Many other names could be added to this list.)

Notes

1. oeis.org.

2. oeis.org/wiki/Works_Citing_OEIS.

3. The OEIS Foundation, Inc., oeisf.org.

4. The circles may have different radii. Two arrangements are considered the same if one can be continuously changed to the other while keeping all circles circular (although the radii may be continuously changed), without changing the multiplicity of intersection points, and without a circle passing through an intersection point. Turning the whole configuration over is allowed.

5. Thanks to Jean-Paul Allouche for this argument.

6. Although in *A Mathematician's Apology*, G. H. Hardy, referring to the fact that 1089 and 2178 are the smallest numbers which when written backward are nontrivial multiples of themselves (**A008919**), remarked that this fact was "likely to amuse amateurs," but was not of interest to mathematicians.

7. In what follows, we tacitly assume $n \geq 2$, to avoid having to repeatedly say "(or 1)."

8. Found by Hans Havermann.

9. mersenneforum.org/showthread.php?t=20527.

10. A superb successor to Felix Klein's 1908, *Elementary Mathematics from an Advanced Standpoint*.

11. Of course the fact that the same number 110 was involved could not possibly be anything more than a coincidence.

12. probabilitysports.com/tilings.html.

13. www.dr-mikes-math-games-for-kids.com/archimedean-graph-paper.html, https://www.printablepaper.net/category/graph.

14. rcsr.net, topospro.com.

15. And more are needed—if interested, please contact the author.

References

[1] D. L. Applegate, H. Havermann, B. Selcoe, V. Shevelev, N. J. A. Sloane, and R. Zum-keller, The Yellowstone Permutation, *J. Integer Seqs.*, **18** (2015), #15.6.7. **MR3360900.**

[2] D. Chavey, Tilings by regular polygons II: A catalog of tilings, *Computers & Mathematics with Applications*, **17:1–3** (1989), 147–165. **MR994197.**

[3] P. Erdős, R. Freud, and N. Hegyvári, Arithmetical properties of permutations of integers, *Acta Math. Hungar.*, **41** (1983), 169–176. **MR704537.**

[4] C. Goodman-Strauss and N. J. A. Sloane, A coloring book approach to finding coordination sequences, *Acta. Cryst. A*, **A75** (2019), 121–134.

[5] R. K. Guy, *Unsolved Problems in Number Theory*, 3rd ed., Springer-Verlag, New York, 2010.

[6] J. C. Lagarias, E. M. Rains, and N. J. A. Sloane, *Experimental Math.*, **11** (2002), 437–446.

[7] M. O'Keefe and B. G. Hyde, Plane nets in crystal chemistry, *Phil. Trans. Royal Soc. London, Series A, Mathematical and Physical Sciences*, **295:1417** (1980), 553–618. MR569155.

[8] J. Stillwell, *Elements of Mathematics: From Euclid to Gödel*, Princeton University Press, Princeton, NJ, 2016.

[9] P. Vanderlind, R. K. Guy, and L. C. Larson, *The Inquisitive Problem Solver*, MAA Press, Washington, DC, 2002. MR1917371.

Mathematics for Big Data

Alessandro Di Bucchianico, Laura Iapichino,
Nelly Litvak, Frank van der Meulen,
and Ron Wehrens

Big data has become a buzzword in the past decade, both in science and among the general public. Scientists from all areas encounter this notion in the shift of content and methods in their research, as well as in current scientific funding programs. For example, big data is one of the selected routes in the Dutch National Scientific Agenda (NWA), and the large funding program Commit2Data has been launched in the Dutch Digital Delta in 2016.

As the Big Data Team of the 4TU Applied Mathematics Institute, we feel that mathematicians should actively engage in big data activities. It is the goal of this article to show the importance of mathematics in big data.

The role of mathematics is easy to overlook and not fully recognized because technological advances are much more visible than mathematical advances, even though the latter often have more effect. Here is a small illustration. It is common knowledge that the acceleration of computers caused by technological advances follows Moore's law: doubling of speed every 18 months. However, it is much less known that the acceleration caused by advances in mathematical methods in scientific computing and optimization is at least of the same order of magnitude, and in some areas even much higher (Bixby 2012, Schilders 2008).

In this essay, we present several explicit real-life examples of the mathematics behind big data, highlighting the role and importance of specific areas of mathematics in these contexts. We show a wide variety of examples: search engines, virtual prototyping in manufacturing, data assimilation, web data analytics, health care, recommendation systems, genomics and other *omics* sciences, and precision farming. In this way, we hope to stimulate mathematicians to work on topics related to big data, as well as to encourage industries and researchers in computer science and other fields to collaborate with mathematicians in this direction.

Similar and more detailed accounts have appeared at other places (Fan et al. 2014, Lindsay et al. 2004, National Research Council 2013, London Workshop on the Future of the Statistical Sciences (2013).

Search Engines

The quality of a search engine depends greatly on ranking algorithms that define in which order web pages appear for the user. This is indeed crucial because most of us do not go beyond the first page of search results. Google's PageRank, at the very heart of the success of Google, was the first and most famous ranking algorithm.

The revolutionary idea of Google was that the importance of a web page depends not only on quantity, but also on quality of links that point to this page. This concept can be seen on a small example from Wikipedia in Figure 1.

The size of the nodes represents their PageRank score. Node B has a large PageRank because it has many incoming links. The PageRank of node C is high because it received the only outgoing link from the important node B. Mathematically, the World Wide Web is modeled as a graph with pages as nodes and hyperlinks as directed edges, and then a large set of equations is solved to find the PageRank values for each node in the graph.

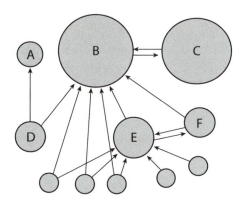

FIGURE 1. PageRank, example from Wikipedia.

PageRank

"Easily bored" surfer. Consider a simple model of a surfer browsing web pages. With probability α, the surfer follows a randomly chosen outgoing link of a page, and with probability $1 - \alpha$ the surfer is bored and jumps to a random page. Initially, Google used $\alpha = 0.85$. PageRank of a page is the stationary (long-run) probability that the surfer visits this page.

Eigenvector. Equivalently, PageRank is the so-called *dominant left eigenvector* of the transition matrix of the above process: the entry (i, j) of this matrix is the probability that the surfer on page i will proceed to page j. Such an eigenvector is unique. The PageRank of a web page is the corresponding component of this unique dominant left eigenvector.

Right after PageRank was introduced, its fast computation became a problem of great interest because the Google matrix is huge; at this moment, it would have hundreds of billions of rows and columns. In the beginning of this century, major speed gains were achieved because of sophisticated new methods from, mainly, linear algebra (Berkhin 2015). Another interesting mathematical and practical problem is the vulnerability of PageRank to deliberate manipulations, such as link farms created intentionally to boost the PageRank.

If we want to predict effectiveness of ranking, it is also important to understand its relation to the network structure. Can we predict the largest PageRank, investigate its stability, or pick up a signal from hidden communities? Can we use ranking to detect important changes in the network structure? A lot of empirical results are available, but they do not answer these questions in sufficient generality. To solve these and other problems, we need to develop new approaches in probability theory and the theory of random graphs (e.g., Chen et al. 2017).

Virtual Prototyping in Manufacturing

High development costs in industry have led many manufacturers to replace building and testing physical prototypes by virtual prototyping, i.e., testing using large-scale simulations of extensive mathematical models based on physical principles. Specific examples are the automotive industry and the aircraft industry (see, e.g., the Virtual Hybrid Testing Framework of Airbus described in Garcia [2013]). Such simulations should be handled with care since there is uncertainty in the outcomes caused by both model limitations and the numerical accuracy of the simulations, often requiring solving large systems of differential equations. There is uncertainty caused by replacing physical reality by a mathematical model. This uncertainty involves both the uncertainty caused by the modeling simplifications (structural uncertainty) and the uncertainty in knowing model parameters (parameter uncertainty). On the other hand, given a complicated mathematical model, it is important to know how accurately numerical methods can approximate specified outputs from this model.

The term *uncertainty quantification* is often used as general term for scientific research in this area. There exist several mathematical approaches to study this uncertainty. One such approach is applying statistical techniques related to experimental design for computer experiments, like Latin hypercube sampling and response surface methods. Another approach is to cast the mathematical model as a stochastic partial differential equation and try to solve that. Recent high-level mathematics combining analysis and stochastics is used, such as perturbation expansion methods for random fields, stochastic operator expansions, and polynomial chaos (Wiener chaos).

Model order reduction (MOR) techniques (e.g., Schilders 2008) have been introduced and exploited to overcome the issue of severe computational times required for solving mathematical models of real-life processes. Over the past four decades, reduced-order models have been developed aimed at replacing the original large-dimension numerical problem (typically called *high-fidelity approximation*) with a reduced problem of substantially smaller dimension. Depending on the context, there are different strategies to generate the reduced problem from the high-fidelity one, e.g., Krylov subspace-based methods, moment matching techniques, proper orthogonal decomposition, balanced truncation, and reduced basis methods. Very short CPU times and limited storage capacities demanded today by MOR methods allow us to tackle a wide range of problems arising in engineering, computational science, and physical and biological sciences.

Data Assimilation

Weather forecasting, for some people the main reason to watch the news, is a data-intensive computational problem with many economic implications (e.g., agriculture, hospitality business, airlines, health care, large public events). The change over time of measurable atmospheric quantities can be described in terms of dynamical systems, transferring information in time-ordered observed data to a physical model of the system. This process is often referred to as data assimilation. Its development has been highly influenced by professionals working in the atmospheric and oceanographic sciences. When discretized in space, a typical model for numerical weather prediction is a differential equation system with dimension of order 10^9 (Law et al. 2015). The state variable of the dynamical system may represent unknown quantities, such as velocity, temperature, and pressure at a grid of locations.

The application of mathematical models to large dynamic data sets has naturally popped up in many other communities as well. Within signal processing, recovering the unknown state of the dynamical system is known as *filtering* or *smoothing*, where the first term refers to online recovery (as opposed to static recovery). Probabilists and statisticians usually speak of *state* and *parameter estimation*. Over the past 30 years, there has been tremendous progress with this type of problems. Under specific assumptions on the dynamical system, computationally efficient methods, such as the (ensemble) Kalman filter, can be used. In more general settings, a Bayesian formulation of the problem and application of Markov Chain Monte Carlo methods and sequential Monte Carlo methods can be exploited (e.g., Robert and Casella 2004, Särkkä 2013). Whereas these methods are currently not yet applicable to weather forecasting, they have proved to be powerful in simplified problems of less demanding dimensions and constitute an active area of research (Cuzol and Memin 2009, Law and Stuart 2012).

Web Data Analytics

Many companies collect large amounts of customer data through their web services. However, having these data does not mean that we already know everything. Even simple tasks, such as counting the number of distinct records in a large customer database (e.g., the number of distinct

HyperLogLog

Hash functions. Each digital object is converted to a sequence of zeros and ones using hash functions. On a set of different objects, a good hash function appears as if randomly generated: zeros and ones have probability $1/2$, independently of each other. The idea of LogLog-type algorithms is to sweep through objects keeping in the memory *only the largest number of zeros* at the beginning hash functions. For example, if we observed

$$00101, 10011, 01010,$$

we remember 2, the largest number of zeros. Roughly, the probability to see 2 zeros followed by one at the beginning of the hash function is

$$\frac{1}{2} \times \frac{1}{2} \times \frac{1}{2} = \frac{1}{8},$$

so we conclude that we saw approximately 8 objects!

HyperLogLog. In this form, the estimation is obviously *too rough*, so it cannot be directly used in practice. A lot of mathematics went into making the result more precise. This math includes dividing hash functions into registers, using different corrections for small and large samples, and harmonic averages. All these ideas are included in HyperLogLog, ensuring its applicability. Further improvements are possible (Heule et al. 2013).

Why LogLog? Assume that we have N objects. Then hash functions have length $\log_2(N)$. Hence, the number of zeros is a number between 0 and $\log_2(N)$, so we need only $\log_2\log_2(N)$ bits of memory to remember this number.

customers that use a certain service), requires advanced mathematics. The exact counting is computationally prohibitive mainly because we cannot keep all objects in the restricted working memory of a computer. However, we might not need that level of accuracy; in such cases, it is often sufficient to work with approximate estimates. Probability theory has been essential in developing algorithms such as Count-Min sketch, MinHash, and HyperLogLog that use random hash functions to store answers. Such algorithms may be accurate within 2% while using only memory in the order of the correct mathematical and/or physical terminology. An important issue in developing these algorithms is to control the variance of the estimators to get consistently accurate estimates.

HyperLogLog is one of the most elegant mathematical solutions for counting distinct objects in big data applications, widely used in practice. Researchers at Google (Heule et al. 2013) state that Google's data analysis system PowerDrill routinely performs about 5 million *count distinct objects* computations per day. In about 100 cases, the resulting number is greater than 1 billion. In 2014, HyperLogLog was implemented by Amazon's data structure store, Redis, as well. An interesting human interest note: the commands of HyperLogLog begin with PF, the initials of the French mathematician Philippe Flajolet, who developed this algorithm (Flajolet et al. 2007).

Maybe even more exciting from a scientific point of view was the result in Backstrom et al. (2012) where HyperLogLog was used to accomplish the incredible task of computing average distances in the complete Facebook graph of more than 700 million nodes. It turned out that the distance (the number of hops along the edges of the Facebook graph) between two Facebook users is on average less than 4!

Health Care

Medical devices such as MRI scanners obtain large-image data at relatively low velocity. Efforts are undertaken to reduce the time it takes to makes scans (typically 30 minutes) since hospitals could obtain higher efficiency with the expensive MRI equipment and patients would suffer less from the unpleasant high noise levels. Making scans at a lower resolution is not an option because of medical reasons. An MRI scan uses magnetic fields to order the spins of hydrogen atoms and radio waves to disturb these spins. When the spins return to their original position, energy is emitted. This energy is measured so that one gets an indication of the amount of tissue is being scanned. Using magnetic gradients, it is possible to localize these measurements.

The mathematical bottom line of this procedure is that MRI scans produce Fourier coefficients one by one. Traditional approaches to reconstruction algorithms cannot yield the desired reduction of scanning time because of the Nyquist–Shannon criterion. Again, advanced mathematical techniques have provided the breakthrough. The basic idea is to project the observed data onto a smaller subspace using sparsity in the data. Remarkably, random projections yield sampling strategies and reconstruction algorithms that outperform traditional signal processing techniques. These methods are known under the name *compressed sensing*. For other applications of compressed sensing in health care, we refer to the Siemens website (2019).

Compressed sensing has been applied successfully in a wide range of other tasks as well, including network tomography, electron microscopy, and facial recognition.

Recommender Systems

Web shops such as Amazon analyze the buying behavior of their customers and present visitors to the Amazon website with recommendations of books and other items that may be of interest. In a similar way, Netflix gives suggestions for movies to its customers. A way to provide such recommendations is to set up a matrix of user ratings of movies (columns are ratings, rows are users). Of course, such a matrix has many empty entries, since there are many more movies (Netflix has around 20,000) than people can see and rate.

The idea behind the recommender systems is that there are relatively few "latent" features that drive our preferences (a sparsity principle). That is, there are a few typical items (books or movies) and a few typical users. Translated into matrices, this situation requires looking for a nonnegative matrix factorization of the preference matrix. This means that a very large and sparse preference matrix is presented as a product of two matrices with much lower dimensions. Although computers become faster, this acceleration is mainly an increase in CPU speed (the speed with which mathematical functions are performed) and much less in faster memory (mainly the speed it takes to find data in memory). Factorizations of large matrices, however, require a huge amount of communication between RAM (working memory) and hard disk ("storage" memory). There is thus a need for memory-efficient factorization algorithms that go far beyond traditional factorization algorithms for singular value decompositions. See Koren et al. (2009) for a technical account by the team that won the $1 million Netflix Prize competition. An exciting new approach in this field is the use of randomized methods, such as stochastic gradient algorithms (Aggarwal 2016).

Genomics and Other Omics Sciences

Now that technology has become available (and affordable!) to rapidly obtain information about the genetic composition of biological samples, huge quantities of data are generated routinely. This is not only true when looking at genetic information (hence, the term *genomics*) but also when looking at proteins (proteomics) and metabolites (metabolomics), to name just two other members of the omics family. The big data aspect here refers to the huge amount of information that we have on a relatively small number of subjects. A typical example is genetic information on humans, animals, or plants that consists of millions of measurements (data points) for each subject. The resulting high-dimensional data require the development of new statistical techniques to draw correct conclusions because traditional statistical methods for such data lead to an unacceptable high number of false positives (Bühlmann and Van de Geer 2013).

Furthermore, advanced data processing methods are needed to convert the measured data into information—one example is the BLAST algorithm (Altschul et al. 1990), incidentally also the most highly cited paper of the 1990s), used to align sequences of nucleotides or amino acids with database entries. In each case, we are confronted with the issue mentioned before: We know an awful lot about few samples, which makes statistical analysis extremely hard. Typical questions are finding genes, proteins, or metabolites related to certain traits or treatment effects. Network analysis is getting more and more attention (Mitra et al. 2013) as a means to bring experimental results into the realm of the things we already know about the biology of the system—one of the main challenges is to combine the different omics data layers into coherent models that explain the behavior of the system under study (Hawkins et al. 2010).

Precision Farming

Agriculture is rapidly becoming a data-rich environment. Tractors are currently connected to the Internet 24/7 and resemble computers on (large) wheels rather than the dusty and primitive muscle machines they were in the twentieth century. As a result, new questions can be addressed that were unthinkable only 10 or 20 years ago. By combining several different information sources (satellite images, plant growth models, management data on plot level), the farmer can, e.g., try to devise optimal strategies to deliver the right amount of water and nutrients to his or her land and in this way obtain the highest possible yield (e.g., Behmann et al. 2015, Clay et al. 2017, and many others).

Here, the problems are the typical big data problems: Even assuming that one has access to all databases and knows how to read and use the data, it is not a trivial question how to combine data with very different characteristics, found in different locations and measured for different purposes. One thing is certain: mathematics and statistics play a pivotal role.

Conclusion

Mathematics and statistics, which are extremely generic tools, have played an important part in technological and scientific developments over the past centuries, and will continue to do so in this big data era. Not only will they contribute to solving problems faster and more efficiently, they will also expand our horizon, exposing questions that we never thought about and maybe did not even expect to be solvable. It is important to realize that advances in this area have both a push and a pull component: Without being confronted with real-life problems, we might lack the incentive or the direction to pursue promising avenues, but without fundamental knowledge we simply lack the tools to tackle the problems successfully. This idea was expressed in a concise way by Bin Yu in her 2014 Institute of Mathematical Statistics presidential address:

> Work on real problems, relevant theory will follow.

Hence, the stress on the applications in this paper: mathematics needs them, just like the applications need mathematics.

References

C. C. Aggarwal, *Recommender Systems*, Cham, Switzerland: Springer, 2016.

S. Altschul, W. Gish, W. Miller, E. Myers, and D. Lipman, Basic local alignment search tool, *Journal of Molecular Biology* 215(3) (1990), 403–410.

L. Backstrom, P. Boldi, M. Rosa, J. Ugander, and S. Vigna, Four degrees of separation, *Proceedings of the 4th Annual ACM Web Science Conference*, ACM: New York, 2012, pp. 33–42.

J. Behmann, A. Mahlein, T. Rumpf, C. Römer, and L. Plümer, A review of advanced machine learning methods for the detection of biotic stress in precision crop protection, *Precision Agriculture* 16 (2015), 239–260.

P. Berkhin, A Survey on PageRank Computing, *Internet Mathematics* 2(1) (2015), 73–120.

R. E. Bixby, A brief history of linear and mixed-integer programming computation, *Documenta Mathematica* (2012), 107–121.

P. Bühlmann and S. A. van de Geer, *Statistics for High-Dimensional Data: Methods, Theory and Applications*, Heidelberg, Germany: Springer, 2013.

A. Cuzol and E. A. Memin, Stochastic filtering technique for fluid flow velocity fields tracking, *IEEE Transactions on Pattern Analysis and Machine Intelligence* 31(7) (2009), 1278–1293.

N. Chen, N. Litvak, and M. Olvera Cravioto, Generalized PageRank on directed configuration networks, *Random Structures & Algorithms* 51(2) (2017), 237–274.

D. E. Clay, S. A. Clay and S. A. Bruggeman, eds., *Practical Mathematics for Precision Farming*, Madison, WI: American Society of Agronomy, Crop Science Society of America, and Soil Science Society of America, 2017.

J. Fan, F. Han, and H. Liu, Challenges of Big Data Analysis, *National Science Review* 1(2) (2014), 293–314.

P. Flajolet, E. Fusy, G. Olivier, and F. Meunier, HyperLogLog: The analysis of a near-optimal cardinality estimation algorithm, in *AofA'07: Proceedings of the 2007 International Conference on Analysis of Algorithms*, Nancy, France: Discrete Mathematics & Theoretical Computer Science (DMTCS), 2007.

E. Garcia, AIRBUS Virtual Hybrid Testing Framework: focus on V&V concerns, presentation at GDR Mascot-Num workshop on Model validation. http://www.gdr-mascotnum .fr/media/vht_garcia.pdf, 2013 (accessed April 16, 2019).

R. D. Hawkins, G. C. Hon, and B. Ren, Next-generation genomics: An integrative approach, *Nature Reviews Genetics* 11 (2010), 476–486.

S. Heule, M. Nunkesser, and A. Hall, Hyper-LogLog in practice: Algorithmic engineering of a state of the art cardinality estimation algorithm, *Proceedings of the 16th International Conference on Extending Database Technology*, ACM: New York, 2013, pp. 683–692.

Y. Koren, R. Bell, and C. Volinsky, Matrix factorization techniques for recommender systems, *Computer* 8 (2009), 42–49.

K. J. H. Law and A. M. Stuart, Evaluating data assimilation algorithms, *Monthly Weather Review* 140 (2012), 3757–3782.

K. J. H. Law, A. M. Stuart, and K. C. Zygalakis, *Data Assimilation: A Mathematical Introduction*, Springer Texts in Applied Mathematics, Vol. 62, Cham, Switzerland: Springer, 2015.

B. G. Lindsay, J. Kettenring, and D. O. Siegmund, A Report on the Future of Statistics, *Statistical Science* 19(3) (2004), 387–413.

London Workshop on the Future of the Statistical Sciences, 2013. http://bit.ly/londonreport (accessed April 16, 2019).

K. Mitra, A. R. Carvunis, S. K. Ramesh, and T. Ideker, Integrative approaches for finding modular structure in biological networks, *Nat. Rev. Genet.* 14 (2013), 719–732.

National Research Council. *Frontiers in Massive Data Analysis*, Washington, DC: National Academies Press, 2013. Freely accessible through https://www.nap.edu/catalog/18374 /frontiers-in-massive-data-analysis (accessed April 16, 2019).

C. P. Robert and G. Casella, *Monte Carlo Statistical Methods*, 2nd ed., Springer Texts in Statistics, New York: Springer, 2004.

S. Särkkä, *Bayesian Filtering and Smoothing*, Cambridge, U.K.: Cambridge University Press, 2013.

W. Schilders, Introduction to Model Order Reduction, in *Mathematics in Industry Model Order Reduction: Theory, Research Aspects and Applications*, Berlin: Springer, 2008, pp. 3–32.

Siemens Healthineers website. https://www.siemens-healthineers.com/magnetic-resonance-imaging/clinical-specialities/compressed-sensing/body-imaging 2019 (accessed April 16, 2019).

B. Yu, Institute of Mathematical Statistics presidential address, *IMS Bulletin* 43, 7 (2014), 1, 13–16. http://bulletin.imstat.org/2014/10/ims-presidential-address-let-us-own-data-science (accessed April 16, 2019).

The Un(solv)able Problem

Toby S. Cubitt, David Pérez-García, and Michael Wolf

The three of us were sitting together in a café in Seefeld, a small town deep in the Austrian Alps. It was the summer of 2012, and we were stuck. Not stuck in the café—the sun was shining, the snow on the Alps was glistening, and the beautiful surroundings were sorely tempting us to abandon the mathematical problem we were stuck on and head outdoors. We were trying to explore the connections between twentieth-century mathematical results by Kurt Gödel and Alan Turing and quantum physics. That, at least, was the dream, a dream that had begun back in 2010, during a semester-long program on quantum information at the Mittag-Leffler Institute near Stockholm.

Some of the questions we were looking into had been explored before by others, but to us this line of research was entirely new, so we were starting with something simple. Just then, we were trying to prove a small and not very significant result to get a feel for things. For months now, we had a proof (of sorts) of this result. But to make the proof work, we had to set up the problem in an artificial and unsatisfying way. It felt like changing the question to suit the answer, and we were not very happy with it. Picking the problem up again during the break after the first session of talks at the workshop in Seefeld that had brought us together in 2012, we still could not see any way around our problems. Half-jokingly, one of us (Michael Wolf) asked, "Why don't we prove the undecidability of something people really care about, like the spectral gap?"

At the time, we were interested in whether certain problems in physics are *decidable* or *undecidable*—that is, can they ever be solved? We had gotten stuck trying to probe the decidability of a much more minor question, one few people care about. The "spectral gap" problem Michael was proposing that we tackle (which we explain later) was one of central importance to physics. We did not know at the time whether this problem was or was not decidable (although we had a hunch it was not) or whether we would be able to prove it either way. But if we could, the results would be of real relevance to physics, not to mention a substantial mathematical achievement. Michael's ambitious suggestion, tossed off almost as a jest, launched us on a grand adventure. Three years and 146 pages of mathematics later, our proof of the undecidability of the spectral gap was published in *Nature*.

To understand what this means, we need to go back to the beginning of the twentieth century and trace some of the threads that gave rise to modern physics, mathematics, and computer science. These disparate ideas all lead back to German mathematician David Hilbert, often regarded as the greatest figure of the past 100 years in the field. (Of course, no one outside of mathematics has heard of him. The discipline is not a good route to fame and celebrity, although it has its own rewards.)

THE SPECTRAL GAP

The authors' mathematical proof took on the question of the *spectral gap*—the jump in energy between the ground state and first excited state of a material. When we think of energy states, we tend to think of electrons in atoms, which can jump up and down between energy levels. Whereas in atoms there is always a gap between such levels, in larger materials made of many atoms, there is sometimes no distance between the ground state and the first excited state: even the smallest possible amount of energy is enough to push the material up an energy level. Such materials are called *gapless*. The authors proved that it will never be possible to determine whether all materials are gapped or gapless. See also color images.

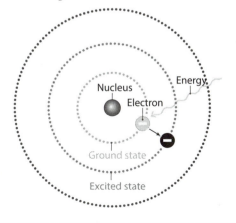

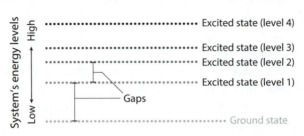

Gapped System. There are discrete gaps between each energy level, and the material must reach a certain energy to make the leap to the next level. See also color images.

Gapless System. No expanse separates the ground state and first excited state, and the material may become excited with just the tiniest input of energy. See also color images.

The Mathematics of Quantum Mechanics

Hilbert's influence on mathematics was immense. Early on, he developed a branch of mathematics called functional analysis—in particular, an area known as spectral theory, which would end up being key to the question within our proof. Hilbert was interested in this area for purely abstract reasons. But as so often happens, his mathematics turned out to be exactly what was necessary to understand a question that was perplexing physicists at the time.

If you heat a substance up, it begins to glow as the atoms in it emit light (hence the phrase "red hot"). The yellow-orange light from sodium street lamps is a good example: sodium atoms predominantly emit light at a wavelength of 590 nanometers, in the yellow part of the visible spectrum. Atoms absorb or release light when electrons within them jump between energy levels, and the precise frequency of that light depends on the energy gap between the levels. The frequencies of light emitted by heated materials thus give us a "map" of the gaps between the atom's different energy levels. Explaining these atomic emissions was one of the problems physicists were wrestling with in the first half of the twentieth century. The question led directly to the development of quantum mechanics, and the mathematics of Hilbert's spectral theory played a prime role.

One of these gaps between quantum energy levels is especially important. The lowest possible energy level of a material is called its *ground state*. This is the level it will sit in when it has no heat. To get a material into its ground state, scientists must cool it down to extremely low temperatures in a laboratory. Then, if the material is to do anything other than sit in its ground state, something must excite it to a higher energy. The easiest way is for it to absorb the smallest amount of energy it can, just enough to take it to the next energy level above the ground state—the first excited state. The energy gap between the ground state and this first excited state is so critical that it is often just called the *spectral gap*.

In some materials, there is a large gap between the ground state and the first excited state. In other materials, the energy levels extend all the way down to the ground state without any gaps at all. Whether a material is gapped or gapless has profound consequences for its behavior at low temperatures. It plays a particularly significant role in quantum phase transitions.

A phase transition happens when a material undergoes a sudden and dramatic change in its properties. We are all familiar with some phase transitions—such as water transforming from its solid form of ice into its liquid form when heated up. But there are more exotic quantum phase transitions that happen even when the temperature is kept extremely low. For example, changing the magnetic field around a material or the pressure it is subjected to can cause an insulator to become a superconductor or cause a solid to become a superfluid.

How can a material go through a phase transition at a temperature of absolute zero (−273.15 degrees Celsius), at which there is no heat at all to provide energy? It comes down to the spectral gap. When the spectral gap disappears—when a material is gapless—the energy needed to reach an excited state becomes zero. The tiniest amount of energy is

Turing Machine

Before modern computers existed, mathematician Alan Turing imagined a hypothetical device called a Turing machine that defined what it meant to compute. The machine reads and performs operations on the symbols written on an infinitely long strip of tape that runs through it. The concept turned out to be central to the authors' proof of the undecidability of the spectral gap problem.

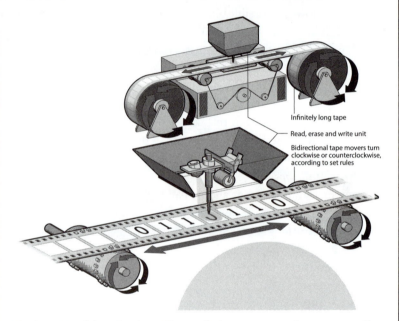

Infinitely long tape

Read, erase and write unit

Bidirectional tape movers turn clockwise or counterclockwise, according to set rules

Turing Machine Basics. The symbols written on the tape initially are the machine's input, and those left on the tape at the end are the answers. The tape can advance or rewind, and the "head" can read, write, or erase the tape's symbols to produce the output.

Halting Problem. Turing devised a simple question known as the *halting problem*: Will a Turing machine running on a given input ever stop? Furthermore, Turing proved that no mathematical procedure could ever answer this question. The authors built on Turing's work to show that the spectral gap is similar to the halting problem and is likewise undecidable.

enough to push the material through a phase transition. In fact, thanks to the weird quantum effects that dominate physics at these very low temperatures, the material can temporarily "borrow" this energy from nowhere, go through a phase transition and "give" the energy back. Therefore, to understand quantum phase transitions and quantum phases, we need to determine when materials are gapped and when they are gapless.

Because this spectral gap problem is so fundamental to understanding quantum phases of matter, it crops up all over the place in theoretical physics. Many famous and long-standing open problems in condensed matter physics boil down to solving this problem for a specific material. A closely related question even crops up in particle physics: there is very good evidence that the fundamental equations describing quarks and their interactions have a *mass gap*. Experimental data from particle colliders, such as the Large Hadron Collider near Geneva, support this notion, as do massive number-crunching results from supercomputers. But proving the idea rigorously from the theory seems to be extremely difficult. So difficult, in fact, that this problem, called the Yang–Mills mass gap problem, has been named one of seven Millennium Prize problems by the Clay Mathematics Institute, and anyone who solves it is entitled to a $1 million prize. All these problems are particular cases of the general spectral gap question. We have bad news for anyone trying to solve them, though. Our proof shows that the general problem is even trickier than we thought. The reason comes down to a question called the *Entscheidungsproblem*.

Unanswerable Questions

By the 1920s, Hilbert had become concerned with putting the foundations of mathematics on a firm, rigorous footing—an endeavor that became known as Hilbert's program. He believed that whatever mathematical conjecture one might make, it is in principle possible to prove either that it is true or that it is false. (It had better not be possible to prove that it is both, or something has gone very wrong with mathematics!) This idea might seem obvious, but mathematics is about establishing concepts with absolute certainty. Hilbert wanted a rigorous proof.

In 1928, he formulated the *Entscheidungsproblem*. Although it sounds like the German sound for a sneeze, in English it translates to "the decision problem." It asks whether there is a procedure, or "algorithm," that can decide whether mathematical statements are true or false.

For example, the statement "Multiplying any whole number by 2 gives an even number" can easily be proved true, using basic logic and arithmetic. Other statements are less clear. What about the following example? "If you take any whole number, and repeatedly multiply it by 3 and add 1 if it's odd, or divide it by 2 if it's even, you always eventually reach the number 1." (Have a think about it.)

Unfortunately for Hilbert, his hopes were to be dashed. In 1931, Gödel published some remarkable results now known as his incompleteness theorems. Gödel showed that there are perfectly reasonable mathematical statements about whole numbers that can be neither proved nor disproved. In a sense, these statements are beyond the reach of logic and arithmetic. And he proved this assertion. If that is hard to wrap your head around, you are in good company. Gödel's incompleteness theorems shook the foundations of mathematics to the core.

Here is a flavor of Gödel's idea: If someone tells you, "This sentence is a lie," is that person telling the truth or lying? If he or she is telling the truth, then the statement must indeed be a lie. But if he or she is lying, then it is true. This quandary is known as the liar paradox. Even though it appears to be a perfectly reasonable English sentence, there is no way to determine whether it is true or false. What Gödel managed to do was to construct a rigorous mathematical version of the liar paradox using only basic arithmetic.

The next major player in the story of the *Entscheidungsproblem* is Alan Turing, the English computer scientist. Turing is most famous among the general public for his role in breaking the German Enigma code during World War II. But among scientists, he is best known for his 1937 paper "On Computable Numbers, with an Application to the *Entscheidungsproblem*." Strongly influenced by Gödel's result, the young Turing had given a negative answer to Hilbert's *Entscheidungsproblem* by proving that no general algorithm to decide whether mathematical statements are true or false can exist. (American mathematician Alonzo Church also independently proved this just before Turing. But Turing's proof was ultimately more significant. Often in mathematics, the proof of a result turns out to be more important than the result itself.)

To solve the *Entscheidungsproblem*, Turing had to pin down precisely what it meant to compute something. Nowadays, we think of computers as electronic devices that sit on our desks, on our laps, or even in our pockets. But computers as we know them did not exist in 1936. In fact, *computer* originally meant a person who carried out calculations with pen and paper. Nevertheless, computing with pen and paper as you did in high school is mathematically no different to computing with a modern desktop computer—just much slower and far more prone to mistakes.

Turing came up with an idealized, imaginary computer called a Turing machine. This very simple imaginary machine does not look like a modern computer, but it can compute everything that the most powerful modern computer can. In fact, any question that can ever be computed (even on quantum computers or computers from the 31st century that have yet to be invented) could also be computed on a Turing machine. It would just take the Turing machine much longer.

A Turing machine has an infinitely long ribbon of tape and a "head" that can read and write one symbol at a time on the tape, then move one step to the right or left along it. The input to the computation is whatever symbols are originally written on the tape, and the output is whatever is left written on it when the Turing machine finally stops running (halts). The invention of the Turing machine was more important even than the solution to the *Entscheidungsproblem*. By giving a precise, mathematically rigorous formulation of what it meant to make a computation, Turing founded the modern field of computer science.

Having constructed his imaginary mathematical model of a computer, Turing then went on to prove that there is a simple question about Turing machines that no mathematical procedure can ever decide: Will a Turing machine running on a given input ever halt? This question is known as the halting problem. At the time, this result was shocking. Nowadays, mathematicians have become accustomed to the fact that any conjecture we are working on could be provable, disprovable, or undecidable.

Where We Come In

In our result, we had to tie all these disparate threads back together. We wanted to unite the quantum mechanics of the spectral gap, the computer science of undecidability, and Hilbert's spectral theory to prove that—like the halting problem—the spectral gap problem was one of the undecidable ones that Gödel and Turing taught us about.

Chatting in that café in Seefeld in 2012, we had an idea for how we might be able to prove a weaker mathematical result related to the spectral gap. We tossed this idea around, not even scribbling on the back of a napkin, and it seemed like it might work. Then the next session of talks started. And there we left it.

A few months later, one of us (Toby Cubitt) visited Michael in Munich, and we did what we had not done in Seefeld: jotted some equations down on a scrap of paper and convinced ourselves that the idea worked. In the following weeks, we completed the argument and wrote it up properly in a private four-page note. (Nothing in mathematics is truly proved until you write it down—or, better still, type it up and show it to a colleague for scrutiny.) Conceptually, this was a major advance. Before now, the idea of proving the undecidability of the spectral gap was more of a joke than a serious prospect. Now we had the first glimmerings that it might actually be possible. But there was still a long way to go. We could not extend our initial idea to prove the undecidability of the spectral gap problem itself.

Burning the Midnight Coffee

We attempted to make the next leap by linking the spectral gap problem to quantum computing. In 1985, Nobel Prize–winning physicist Richard Feynman published one of the papers that launched the idea of quantum computers. In that paper, Feynman showed how to relate ground states of quantum systems to computation. Computation is a dynamic process: you supply the computer with input, and it goes through several steps to compute a result and outputs the answer. But ground states of quantum systems are completely static: the ground state is just the configuration a material sits in at zero temperature, doing nothing at all. So how can it make a computation?

The answer comes through one of the defining features of quantum mechanics: *superposition*, which is the ability of objects to occupy many states simultaneously, as, for instance, Erwin Schrödinger's famous quantum cat, which can be alive and dead at the same time. Feynman proposed constructing a quantum state that is in a superposition of the various steps in a computation—initial input, every intermediate step of the computation, and final output—all at once. Alexei Kitaev of the California Institute of Technology later developed this idea substantially by constructing an imaginary quantum material whose ground state looks exactly like this.

If we used Kitaev's construction to put the entire history of a Turing machine into the material's ground state in superposition, could we transform the halting problem into the spectral gap problem? In other words, could we show that any method for solving the spectral gap problem would also solve the halting problem? Because Turing had already shown that the halting problem was undecidable, this step would prove that the spectral gap problem must also be undecidable.

Encoding the halting problem in a quantum state was not a new idea. Seth Lloyd, now at the Massachusetts Institute of Technology, had proposed this almost two decades earlier to show the undecidability of another quantum question. Daniel Gottesman of the Perimeter Institute for Theoretical Physics in Waterloo and Sandy Irani of the University of California, Irvine, had used Kitaev's idea to prove that even single lines of interacting quantum particles can show very complex behavior. In fact, it was Gottesman and Irani's version of Kitaev's construction that we hoped to make use of.

But the spectral gap is a different kind of problem, and we faced some apparently insurmountable mathematical obstacles. The first had to do with supplying the input into the Turing machine. Remember that the undecidability of the halting problem is about whether the Turing machine halts *on a given input*. How could we design our imaginary quantum material in a way that would let us choose the input to the Turing machine to be encoded in the ground state?

When working on that earlier problem (the one we were still stuck on in the café in Seefeld), we had an idea of how to rectify the issue by putting a "twist" in the interactions between the particles and using the angle of this rotation to create an input to the Turing machine. In January 2013, we met at a conference in Beijing and discussed this plan together. But we quickly realized that what we had to prove came very close to contradicting known results about quantum Turing machines. We decided that we needed a complete and rigorous proof that our idea worked before we pursued the project further.

At this point, Toby had been part of David Pérez-García's group at Complutense University of Madrid for more than two years. In that same month, he moved to the University of Cambridge, but his new apartment there was not yet ready, so his friend and fellow quantum information theorist Ashley Montanaro offered to put him up. For those

TILING AN INFINITE BATHROOM FLOOR

To connect the spectral gap problem to the halting problem, the authors considered the classic mathematical question of how to tile an infinitely large floor. Imagine that you have a box with a certain selection of tiles, and you want to arrange them so that the colors on the sides of each tile match those next to them. In some cases, this is possible by tiling the floor in either a repeating periodic pattern or a fractallike aperiodic pattern.

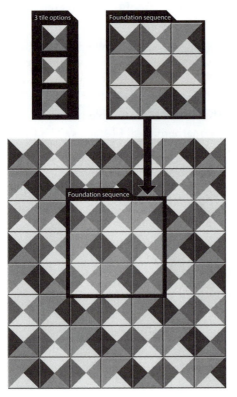

Periodic Tiles. One version of the classic problem concerns tiles that come in three varieties containing five colors. In this particular case, it is possible to tile the floor with all sides matching up by creating a rectangle that repeats. On each side of the rectangle, the colors match so that many versions of the same rectangle can be placed next to one another in an infinite pattern. See also color images.

two months, he set to work producing a rigorous proof of this idea. His friend would find him at the kitchen table in the morning, a row of empty coffee mugs next to him, about to head to bed, having worked through the night figuring out details and typing them up. At the end of those two months, Toby sent around the completed proof.

Aperiodic Tiles

In their proof, the authors used a particular set of tiles designed by mathematician Rafael Robinson in 1971. Robinson's tiles fit together in an ever-expanding sequence that does not quite repeat but instead creates a fractal-like pattern. All rotations of the six tiles shown here are allowed. There are also other ways to fit these pieces together in a periodic pattern, but by adding more markings to these tiles (*not shown*), Robinson designed a set of 56 tiles for which no pattern is possible other than the one shown.

6 tile options

In Remembrance of Tilings Past

This 29-page proof showed how to overcome one of the obstacles to connecting the ground state of a quantum material to computation with a Turing machine. But there was an even bigger obstacle to that goal: the resulting quantum material was always gapless. If it is always gapless, the spectral gap problem for this particular material is easy to solve: the answer is gapless!

Our first idea from Seefeld, which proved a much weaker result than we wanted, nonetheless managed to get around this obstacle. The key was using *tilings*. Imagine that you are covering a large bathroom floor with tiles. In fact, imagine that it is an infinitely big bathroom. The tiles have a simple pattern on them: each of the four sides of the tile is a different color. You have various boxes of tiles, each with a different arrangement of colors. Now imagine that there is an infinite supply of tiles in each box. You, of course, want to tile the infinite bathroom floor so that the colors on adjacent tiles match. Is this possible?

The answer depends on which boxes of tiles you have available. With some sets of colored tiles, you will be able to tile the infinite bathroom floor. With others, you will not. Before you select which boxes of tiles to buy, you would like to know whether or not they will work. Unfortunately for you, in 1966 mathematician Robert Berger proved that this problem is undecidable.

One easy way to tile the infinite bathroom floor would be to first tile a small rectangle so that colors on opposite sides of it match. You could then cover the entire floor by repeating this rectangular pattern. Because they repeat every few tiles, such patterns are called *periodic*. The reason the tiling problem is undecidable is that nonperiodic tilings also exist: patterns that cover the infinite floor but never repeat.

Back when we were discussing our first small result, we studied a 1971 simplification of Berger's original proof made by Rafael Robinson of the University of California, Berkeley. Robinson constructed a set of 56 different boxes of tiles that, when used to tile the floor, produce an interlocking pattern of ever larger squares. This fractal pattern looks periodic, but in fact, it never quite repeats itself. We extensively discussed ways of using tiling results to prove the undecidability of quantum properties. But back then, we were not even thinking about the spectral gap. The idea lay dormant.

In April 2013, Toby paid a visit to Charlie Bennett at IBM's Thomas J. Watson Research Center. Among Bennett's many achievements before becoming one of the founding fathers of quantum information theory was his seminal 1970s work on Turing machines. We wanted to quiz him about some technical details of our proof to make sure that we were not overlooking something. He said that he had not thought about this stuff for 40 years, and it was high time a younger generation took over. (He then went on to helpfully explain some subtle mathematical details of his 1970s work, which reassured us that our proof was okay.)

Bennett has an immense store of scientific knowledge. Because we had been talking about Turing machines and undecidability, he e-mailed copies of a couple of old papers on undecidability that he thought might interest us. One of these was the same 1971 paper by Robinson that we had studied. Now the time was right for the ideas sowed in our earlier discussions to spring to life. Reading Robinson's paper again, we realized that it was exactly what we needed to prevent the spectral gap from vanishing.

Our initial idea had been to encode one copy of the Turing machine into the ground state. By carefully designing the interactions between the particles, we could make the ground state energy a bit higher if the Turing machine halted. The spectral gap—the energy jump to the first excited state—would then depend on whether the Turing machine halted or not. There was just one problem with this idea, and it was a big one. As the number of particles increased, the additional contribution to the ground state energy got closer and closer to zero, leading to a material that was always gapless.

But by adapting Berger's tiling construction, we could instead encode *many copies* of exactly the same Turing machine into the ground state. In fact, we could attach one copy to each square in Robinson's tiling pattern. Because these are identical copies of the same Turing machine, if one of them halts, they all halt. The energy contributions from all these copies add up. As the number of particles increases, the number of squares in the tiling pattern grows. Thus, the number of copies of the Turing machine increases, and their energy contribution becomes huge, giving us the possibility of a spectral gap.

Exams and Deadlines

One significant weakness remained in the result we had proved. We could not say anything about how *big* the energy gap was when the material was gapped. This uncertainty left our result open to the criticism that the gap could be so small that it might as well not exist. We needed to prove that the gap, when it existed, was actually large. The first solution we found arose by considering materials in three dimensions instead of the planar materials we had been thinking about until then.

When you cannot stop thinking about a mathematical problem, you make progress in the most unexpected places. David worked on the details of this idea in his head while he was supervising an exam. Walking along the rows of tables in the hall, he was totally oblivious to the students working feverishly around him. Once the test was over, he committed this part of the proof to paper.

We now knew that getting a big spectral gap was possible. Could we also get it in two dimensions, or were three necessary? Remember the problem of tiling an infinite bathroom floor. What we needed to show was that for the Robinson tiling, if you got one tile wrong somewhere, but the colors still matched everywhere else, then the pattern formed by the tiles would be disrupted only in a small region centered on that wrong tile. If we could show this "robustness" of the Robinson tiling, it would imply that there was no way of getting a small spectral gap by breaking the tiling only a tiny bit.

By the late summer of 2013, we felt that we had all the ingredients for our proof to work. But there were still some big details to be resolved, such as proving that the tiling robustness could be merged with all the other proof ingredients to give the complete result. The Isaac Newton Institute for Mathematical Science in Cambridge, England, was hosting a special workshop on quantum information for the whole of the autumn semester of 2013. All three of us were invited to attend. It was the perfect opportunity to work together on finishing the project. But David was not able to stay in Cambridge for long. We were determined to complete the proof before he left.

The Isaac Newton Institute has blackboards everywhere—even in the bathrooms! We chose one of the blackboards in a corridor (the closest to the coffee machine) for our discussions. We spent long hours at the blackboard developing the missing ideas, then divided the task of making these ideas mathematically rigorous among us. This process always takes far more time and effort than it seems on the blackboard. As the date of David's departure loomed, we worked without interruption all day and most of the night. Just a few hours before he left for home, we finally had a complete proof.

In physics and mathematics, researchers make most results public for the first time by posting a draft paper to the arXiv.org preprint server before submitting it to a journal for peer review. Although we were now fairly confident that the entire argument worked and the hardest part was behind us, our proof was not ready to be posted. There were many mathematical details to be filled in. We also wanted to rewrite and tidy up the paper (we hoped to reduce the page count in the process, although in this we would completely fail). Most important, although at least one of us had checked every part of the proof, no one had gone through it all from beginning to end.

In summer 2014, David was on a sabbatical at the Technical University of Munich with Michael. Toby went out to join them. The plan was to spend this time checking and completing the whole proof, line by line. David and Toby were sharing an office. Each morning, David would arrive with a new printout of the draft paper, copious notes and questions scribbled in the margins and on interleaved sheets. The three of us would get coffee and then pick up where we had left off the day before, discussing the next section of the proof at the blackboard. In the afternoon, we divided up the work of rewriting the paper and adding the new material and of going through the next section of the proof. Toby was suffering from a slipped disc and could not sit down, so he worked with his laptop propped on top of an upturned garbage bin on top of the desk. David sat opposite, the growing pile of printouts and notes taking up more and more of his desk. On a couple of occasions, we found significant gaps in the proof. These gaps turned out to be surmountable, but bridging them meant adding substantial material to the proof. The page count continued to grow.

After six weeks, we had checked, completed, and improved every single line of the proof. It would take another six months to finish writing everything up. Finally, in February 2015, we uploaded the paper to arXiv.org.

What It All Means

Ultimately, what do these 146 pages of complicated mathematics tell us?

First, and most important, they give a rigorous mathematical proof that one of the basic questions of quantum physics cannot be solved in general. Note that the "in general" here is critical. Even though the halting problem is undecidable in general, for *particular* inputs to a Turing machine, it is often still possible to say whether it will halt or not. For example, if the first instruction of the input is "halt," the answer is pretty clear. The same goes if the first instruction tells the Turing machine to loop forever. Thus, although undecidability implies that the spectral gap problem cannot be solved for *all* materials, it is entirely possible to solve it for specific materials. In fact, condensed matter physics is littered with such examples. Nevertheless, our result proves rigorously that even a perfect, complete description of the microscopic interactions between a material's particles is not always enough to deduce its macroscopic properties.

You may be asking yourself if this finding has any implications for "real physics." After all, scientists can always try to measure the spectral gap in experiments. Imagine if we could engineer the quantum material from our mathematical proof and produce a piece of it in the lab. Its interactions are so extraordinarily complicated that this task is far, far beyond anything scientists are ever likely to be able to do. But if we could and then took a piece of it and tried to measure its spectral gap, the material could not simply throw up its hands and say, "I can't tell you—it's undecidable." The experiment would have to measure *something*.

The answer to this apparent paradox lies in the fact that, strictly speaking, the terms "gapped" and "gapless" only make mathematical sense when the piece of material is infinitely large. Now, the 10^{23} or so atoms contained in even a very small piece of material represent a very large number indeed. For normal materials, this is close enough to infinity to make no difference. But for the very strange material constructed in our proof, large is not equivalent to infinite. Perhaps with 10^{23} atoms, the material appears in experiments to be gapless. Just to be sure, you take a sample of material twice the size and measure again. Still gapless. Then, late one night, your graduate student comes into the lab and adds just one extra atom. The next morning, when you measure it again, the material has become gapped! Our result proves that the size at which this transition may occur is incomputable (in the same Gödel–Turing sense that you are now familiar with). This story is completely hypothetical for now because we cannot engineer a material this complex. But it shows, backed by a rigorous mathematical proof, that scientists must take special care when extrapolating experimental results to infer the behavior of the same material at larger sizes.

And now we come back to the Yang–Mills problem—the question of whether the equations describing quarks and their interactions have a mass gap. Computer simulations indicate that the answer is yes, but our result suggests that determining for sure may be another matter. Could it be that the computer-simulation evidence for the Yang–Mills mass gap would vanish if we made the simulation just a tiny bit larger? Our result cannot say, but it does open the door to the intriguing possibility that the Yang–Mills problem, and other problems important to physicists, may be undecidable.

And what of that original small and not very significant result we were trying to prove all those years ago in a café in the Austrian Alps? Actually, we are still working on it.

More to Explore

Undecidability and Nonperiodicity for Tilings of the Plane. Raphael M. Robinson in *Inventiones Mathematicae*, Vol. 12, No. 3, pp. 177–209; September 1971.

Undecidability of the Spectral Gap. Toby S. Cubitt, David Pérez-García, and Michael M. Wolf in *Nature*, Vol. 528, pp. 207–211; December 10, 2015. Preprint available at https://arxiv.org/abs/1502.04573.

The Mechanization of Mathematics

Jeremy Avigad

Introduction

In 1998, Thomas Hales announced a proof of the Kepler conjecture, which states that no nonoverlapping arrangement of equal-sized spheres in space can attain a density greater than that achieved by the naive packing obtained by arranging them in nested hexagonal layers. The result relied on extensive computation to enumerate certain combinatorial configurations known as "tame graphs" and to establish hundreds of nonlinear inequalities.

He submitted the result to the *Annals of Mathematics*, which assigned a team of referees to review it. Hales found the process unsatisfying: it was more than four years before the referees began their work in earnest, and they cautioned that they did not have the resources to review the body of code and vouch for its correctness. In response, he launched an effort to develop a formal proof in which every calculation, and every inference, would be fully checked by a computer. To name the project, Hales searched for a word containing the initial letters of the words "formal," "proof," and "Kepler" and settled on "Flyspeck," which means "to scrutinize, or examine carefully." The project was completed in August 2014.[1]

In May 2016, three computer scientists, Marijn Heule, Oliver Kullmann, and Victor Marek, announced a solution to an open problem posed by Ronald Graham. Graham had asked whether it is possible to color the positive integers red and blue in such a way that there are no monochromatic Pythagorean triples, that is, no monochromatic triple a, b, c satisfying $a^2 + b^2 = c^2$. Heule, Kullmann, and Marek determined that it is possible to color the integers from 1 to 7,824 in such a way (Figure 1), but that there is no coloring of the integers from 1 to 7,825

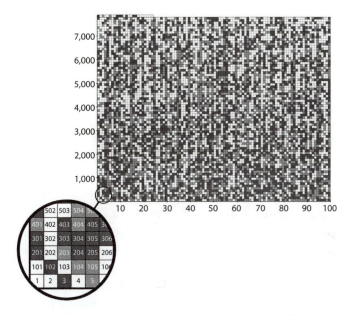

FIGURE 1. A family of colorings of the integers from 1 to 7,824 with no monochromatic Pythagorean triple. White squares can be colored either red or blue. Courtesy of Marijn J. Heule. See also color images.

with this property. They obtained this result by designing, for each n, a propositional formula that describes a coloring of $1, \ldots, n$ with no monochromatic triple. They then used a propositional satisfiability solver, together with heuristics tailored to the particular problem, to search for satisfying assignments for specific values of n.

For $n = 7,824$, the search was successful, yielding an explicit coloring of the corresponding range of integers. For the negative result, however, it is riskier to take the software's failure to find a coloring as an ironclad proof that there isn't one. Instead, Heule, Kullmann, and Marek developed an efficient format to encode a proof that the search was indeed exhaustive, providing a certificate that could be checked by independent means. The resulting proof is 200 terabytes long, leading to popular reports in the international press of the longest proof ever found. They managed to produce a 68-gigabyte certificate with enough information for users to reproduce the proof on their own and made it publicly available.

The use of computers in mathematics is by no means new. Numerical methods are routinely used to predict the weather, model the economy, and track climate change, as well as to make decisions and optimize outcomes in industry. Computer algebra systems like Mathematica, Maple, and Sage are widely used in applied mathematics and engineering.

By now we have even gotten used to the fact that computers can contribute to results in pure mathematics. The 1976 proof of the four-color theorem by Kenneth Appel and Wolfgang Haken used the computer to check that each of a list of 1,936 maps had a required property, and to date there is no proof that can be checked by hand. In 2002, Warwick Tucker used careful calculation to show that the Lorenz attractor exists, that is, that Lorenz's original equations do indeed give rise to chaotic behavior in a precise sense. In doing so, he settled the fourteenth problem on a list of open problems prepared by Stephen Smale at the turn of the twenty-first century. In 2005, Manjul Bhargava and Jonathan Hanke used sophisticated computations to prove a conjecture by John Conway, now called the 290 theorem, which asserts that any positive definite quadratic form with integral coefficients that represents all positive integers up to 290 in fact represents all the positive integers. In 2013, Marc Lackenby and Rob Meyerhoff relied on computer assistance (as well as Perelman's proof of the geometrization conjecture) to provide a sharp bound on exceptional slopes in Thurston's Dehn surgery theorem. Other examples can be found under the Wikipedia entry for "computer-assisted proof," and in a survey by Hales (2014).

But the uses of computation in the Flyspeck project and the solution to the Pythagorean triples problem have a different and less familiar character. Hales' 1998 result was a computer-assisted proof in the conventional sense, but the Flyspeck project was dedicated to *verification*, using the computer to check not only the calculations but also the pen-and-paper components of the proof, including all the background theories, down to constructions of the integers and real numbers. In the work on the Pythagorean triples problem, the computer was used to carry out a heuristic search rather than a directed computation. Moreover, in the negative case, the result of the computation was a formal proof that could be used to certify the correctness of the result.

What these two examples have in common is that they are mathematical instances of what computer scientists refer to as *formal methods*:

computational methods that rely on formal logic to make mathematical assertions, specify and search for objects of interest, and verify results. In particular, both Flyspeck and the Pythagorean triples result rely crucially on formal representations of mathematical assertions and formal notions of mathematical inference and proof.

The thesis I will put forth in this article is that these two results are not isolated curiosities, but, rather, early signs of a fundamental expansion of our capacities for discovering, verifying, and communicating mathematical knowledge. The goal of this article is to provide some historical context, survey the incipient technologies, and assess their long-term prospects.

The Origins of Mechanized Reasoning

Computer scientists, especially those working in automated reasoning and related fields, find a patron saint in Ramon Llull, a thirteenth-century Franciscan monk from Mallorca. Llull is best known for his *Ars generalis ultima* ("ultimate general art"), a work that presents logical and visual aids designed to support reasoning that could win Muslims over to the Christian faith. For example, Llull listed sixteen of God's attributes—goodness, greatness, wisdom, perfection, eternity, and so on—and assigned a letter to each. He then designed three concentric paper circles, each of which had the corresponding letters inscribed around its border. By rotating the circles, one could form all combinations of the three letters, and thereby appreciate the multiplicity of God's attributes (Figure 2). Other devices supported reasoning about the faculties and acts of the soul, the virtues and the vices, and so on.

Although this work sounds quirky today, it is based on three fundamental assumptions that are now so ingrained in our thought that it is hard to appreciate their significance:

- We can represent concepts, assertions, or objects of thought with symbolic tokens.
- Compound concepts (or assertions or thoughts) can be obtained by forming combinations of more basic ones.
- Mechanical devices, even as simple as a series of concentric wheels, can be helpful in constructing and reasoning about such combinations.

FIGURE 2. A thirteenth-century Franciscan monk, Ramon Llull, designed logical and visual aids to reason about the multiplicity of God's attributes. Courtesy of Smithsonian Science Images.

Llull was influenced by an early Muslim thinker, al-Ghazali, and the first two assumptions can be found even earlier in the work of Aristotle. For example, the theory of the syllogism in *Prior Analytics* offers general arguments in which letters stand for arbitrary predicates, and Aristotle's other writings address the question of how predicates can combine to characterize or define a subject. But Llull's use of mechanical devices and procedures to support reasoning was new, and, in the eyes of many, this makes him the founder of mechanized reasoning.

Almost 400 years later, Llull's ideas were an inspiration to Gottfried Leibniz, who, in his doctoral dissertation, dubbed the method *ars combinatoria* ("the art of combinations"). In 1666, he wrote a treatise, *Dissertatio de arte combinatoria*, which contained a mixture of logic and modern combinatorics. The unifying theme once again was a method for combining concepts and reasoning about these combinations. In this treatise, Leibniz famously proposed the development of a *characteristica universalis*, a symbolic language that could express any rational thought, and a *calculus ratiocinator*, a mechanical method for assessing its truth.

Although Leibniz made some initial progress toward this goal, his languages and calculi covered a restricted fragment of logical inference. It is essentially the fragment we now call *propositional logic*, rediscovered by George Boole in the middle of the nineteenth century. But soon after Boole, others began to make good on Leibniz's promise of a universal language of thought, or, at least, languages that were sufficient to represent more complex assertions. Peirce, Schröder, Frege, Peano, and others expanded logical symbolism to include quantifiers and relations. In 1879, Gottlob Frege published his landmark work, *Begriffsschrift* ("concept writing"), which presented an expressive logical language together with axioms and rules of inference. In the introduction, he situated the project clearly in the Leibnizian tradition while carefully restricting its scope to scientific language and reasoning.

In the early twentieth century, the work of David Hilbert and his students and collaborators, Ernst Zermelo's axiomatization of set theory, and Bertrand Russell and Alfred North Whitehead's *Principia Mathematica* all furthered the project of using symbolic systems to provide a foundation for mathematical reasoning. The project was so successful that, in 1931, Kurt Gödel could motivate his incompleteness theorems with the following assessment:

> The development of mathematics toward greater precision has led, as is well known, to the formalization of large tracts of it, so that one can prove any theorem using nothing but a few mechanical rules. The most comprehensive formal systems that have been set up hitherto are the system of *Principia mathematica* (PM) on the one hand and the Zermelo-Fraenkel axiom system of set theory (further developed by J. von Neumann) on the other. These two systems are so comprehensive that in them all methods of proof used today in mathematics are formalized, that is, reduced to a few axioms and rules of inference.

This brief historical overview helps situate the work I intend to present here. To properly bridge the gap from the beginning of the twentieth century to the present, I would have to survey not only the history of logic, foundations of mathematics, and computer science but also the history of automated reasoning and interactive theorem proving. Nothing I can do in the scope of this article would do these subjects justice, so I now set them aside and jump abruptly to the present day.

Formal Methods in Computer Science

The phrase *formal methods* is used to describe a body of methods in computer science for specifying, developing, and verifying complex hardware and software systems. The word *formal* indicates the use of formal languages to write assertions, define objects, and specify constraints. It also indicates the use of formal semantics, that is, accounts of the meaning of a syntactic expression, which can be used to specify the desired behavior of a system or the properties of an object sought. For example, an algorithm may be expected to return a tuple of numbers satisfying a given constraint, C, expressed in some specified language, whereby the logical account spells out what it *means* for an object to satisfy the symbolically expressed constraint. Finally, the word *formal* suggests the use of formal rules of inference, which can be used to verify claims or guide a search.

Put briefly, formal methods are used in computer science to say things, find things, and check things. Using an approach known as *model checking*, an engineer describes a piece of hardware or software and specifies a property that it should satisfy. A tool like a satisfiability solver (SAT solver) or satisfiability-modulo-theories solver (SMT solver) then searches for a counterexample trace, that is, an execution path that violates the specification. The search is designed to be exhaustive so that failure to find such a trace guarantees that the specification holds. In a complementary approach known as *interactive theorem proving*, the engineer seeks to construct, with the help of the computer, a fully detailed formal proof that the artifact meets its specification.

It should not be surprising that such technologies bear on mathematical activity as well. Proving the correctness of a piece of hardware or software is an instance of proving a theorem, in this case, the theorem that states that the hardware or software, described in mathematical terms, meets its specification. Searching for bugs in hardware or software is simply an instance of searching for a mathematical object that satisfies given constraints. Moreover, claims about the behavior of hardware and software are made with respect to a body of mathematical background. For example, verifying software often depends on integer or floating point arithmetic and on properties of basic combinatorial structures. Verifying a hardware control system may invoke properties of dynamical systems, differential equations, and stochastic processes.

Of course, there is a difference in character between proving ordinary mathematical theorems and proving hardware and software correct. Verification problems in computer science are generally difficult because of the volume of detail, but they typically do not have the conceptual depth one finds in mathematical proofs. But although the focus here is on mathematics, you should keep in mind that there is no sharp line between mathematical and computational uses of formal methods, and many of the systems and tools I describe can be used for both purposes.

Verified Proof

Interactive theorem proving involves the use of computational proof assistants to construct formal proofs of mathematical claims using the axioms and rules of a formal foundation that is implemented by the system. The user of such an assistant generally has a proof in mind and works interactively with the system to transform it into a formal derivation. Proofs are presented to the system using a specialized proof language, much like a programming language. The computational assistant processes the input, complains about the parts it cannot understand, keeps track of goals and proof obligations, and responds to queries, say, about definitions and theorems in the background libraries. Most importantly, every inference is checked for correctness using a small, trusted body of code, known as the *kernel* or *trusted computing base*. Some systems even retain, in memory, a complete description of the resulting axiomatic derivation, a complex piece of data that can be exported and verified by an independent reference checker.

The choice of axiomatic foundation varies. Some systems are based on set theory, in which every object denotes a set. Predicates are then used to pick out which sets represent objects like integers, real numbers, functions, triangles, and structures. Most systems, however, implement frameworks in which every object is assigned a *type* that indicates its intended use. For example, an object of type `int` is an integer, and an object of type `int → int` is a function from integers to integers. Such an approach often permits more convenient forms of input, since a system can use knowledge of data types to work out the meaning of a given expression. It also makes it possible for a system to catch straightforward errors, such as when a user applies a function to an object of the wrong type. The complexity of the typing system can vary, however. Some versions of type theory have a natural computational interpretation, so that the definition of a function like the factorial function on the nonnegative integers comes with a means of evaluating it.

Many core theorems of mathematics have been formalized in such systems, such as the prime number theorem, the four-color theorem, the Jordan curve theorem, Gödel's first and second incompleteness theorems, Dirichlet's theorem on primes in an arithmetic progression, the Cartan fixed-point theorems, and the central limit theorem. Verifying a big-name theorem is always satisfying, but a more important measure of progress lies in the mathematical libraries that support them. To date, a substantial body of definitions and theorems from undergraduate mathematics has been formalized, and there are good libraries for elementary number theory, real and complex analysis, point-set topology, measure-theoretic probability, abstract algebra, Galois theory, and so on. In November 2008, the *Notices* devoted a special issue to the topic of interactive theorem proving, which provides an overview of the state of the field at the time (see also Avigad and Harrison 2014). As a result, here I discuss only a few landmarks that have been achieved since then.

In 2012, Georges Gonthier and thirteen coauthors announced the culmination of a six-year project that resulted in the verification of the Feit–Thompson odd order theorem. Feit and Thompson's journal publication in 1963 ran 255 pages, a length that is not shocking by today's standards but was practically unheard of at the time. The formalization was carried out in Coq, a theorem prover based on a constructive type theory using a proof language designed by Gonthier known as SSReflect. The formalization included substantial libraries for finite group theory, linear algebra, and representation theory. All told, the proof comprised roughly 150,000 lines of formal proof, including 4,000 definitions and 13,000 lemmas and theorems.

Another major landmark is the completion of the formal verification of the Kepler conjecture, described in the introduction. Most of the proof was carried out in a theorem prover known as HOL light, though one component, the enumeration of tame graphs, was carried out in Isabelle.

Yet another interesting development in the past few years stems from the realization, due to Steve Awodey and Michael Warren and, independently, Vladimir Voevodsky, that dependent type theory, the logical framework used by a number of interactive theorem provers, has a novel topological interpretation. In this interpretation, data types correspond to topological spaces or, more precisely, abstract representations of topological spaces up to homotopy. Expressions that would ordinarily be understood as functions between data types are interpreted instead as continuous maps. An expression of the form $x = y$ is interpreted as saying that there is a path between x and y, and the rules for reasoning about equality in dependent type theory correspond to a common pattern of reasoning in homotopy theory in which paths are contracted down to a base point. This interpretation opens up possibilities for using interactive theorem provers to reason about subtle topological constructions. Moreover, Voevodsky showed that one can consistently add an axiom that states, roughly, that isomorphic structures are equal, which is to say, the entire language of dependent type theory respects homotopic equivalence. The field has come to be known as *homotopy type theory*, a play on the homotopical intepretation of type theory and the theory of homotopy types.

At this stage, it may seem premature to predict that formally verified proof will become common practice. Even the most striking successes in formally verified mathematics so far have done little to alter the status quo. Hales' result was published in the *Annals of Mathematics* and widely celebrated long before the formal verification was complete, and even though the verification of the Feit–Thompson theorem turned up minor misstatements and gaps in the presentations they followed, the correctness of the theorem was not in doubt, and the repairs were routine.

But the mathematical literature is filled with errors, ranging from typographical errors, missing hypotheses, and overlooked cases to mistakes that invalidate a substantial result. In a talk delivered in 2014,[2] Vladimir Voevodsky surveyed a number of substantial errors in the literature in homotopy theory and higher category theory, including a counterexample, discovered by Carlos Simpson in 1998, to the main result of a paper he himself had published with Michal Kapronov in 1989. Voevodsky ultimately turned to formal verification because he felt that it was necessary for the level of rigor and precision the subject requires.

The situation will only get worse as proofs become longer and more complex. In a 2008 opinion piece in the *Notices*, "Desperately Seeking Mathematical Truth," Melvyn Nathanson lamented the difficulties in certifying mathematical results: "We mathematicians like to talk about the 'reliability' of our literature, but it is, in fact, unreliable." His essay was not meant to be an advertisement for formal verification, but it can easily be read that way.

Checking the details of a mathematical proof is far less enjoyable than exploring new concepts and ideas, but it is important nonetheless.

Rigor is essential to mathematics, and even minor errors are a nuisance to those trying to read, reconstruct, and use mathematical results. Even expository gaps are frustrating, and it would be nice if we could interactively query proofs for more detail, spelling out any inferences that are not obvious to us at first. It seems inevitable that, in the long run, formal methods will deliver such functionality.

Verified Computation

When Hales submitted his proof of the Kepler conjecture to the *Annals*, a sticking point was that the mathematically trained referees were not equipped to vouch for the correctness of the code. Hales and his collaborators countered this concern by verifying these computations as well as the conventional mathematical arguments. This was not the first example of a formally verified proof that involved substantial computation: Gonthier's verification of the four-color theorem in Coq was of a similar nature, relying on a simplified computational approach by Robertson et al. (1996).

This brings us to the subtle question as to what exactly it means to verify a computation. Researchers working in formal verification are sensitive to the question as to what components of a system have to be trusted to ensure the correctness of a result. Ordinary pen-and-paper proofs are checked with respect to the axioms and rules of a foundational deductive system. In that case, the trust lies with the kernel, typically a small, carefully written body of code, as well as the soundness of the axiomatic system itself, the hardware that runs the kernel, and so on. To verify the nonlinear inequalities in the Flyspeck project, Hales and a student of his, Alexey Solovyev, reworked the algorithms so that they produce proofs as they go. Whenever a calculation depended on a fact like $12 \times 7 = 84$, the algorithm would produce a formal proof, which was then checked by the kernel. In other words, every computational claim was subjected to the same standard as a pen-and-paper proof. Checking the nonlinear inequalities involved verifying floating point calculations, and the full process required roughly 5,000 processor hours on the Microsoft Azure cloud.

Another approach to verifying computation involves describing a function in the formal foundational language of a theorem prover, proving that the description meets the desired specification, and then using an automated procedure to extract a program in a conventional programming language to compute its values. The target of the extraction procedure is often a functional programming language like ML or Haskell. This approach requires a higher degree of trust, since it requires that the extraction process preserve the semantics of the formal expression. Of course, one also has to trust the target programming language and its compiler or interpreter. Even so, the verification process imposes a much higher standard of correctness than unverified code. When writing ordinary mathematical code, it is easy to make mistakes, such as omitting corner cases or misjudging the properties that are maintained by an iterative loop. In the approach just described, every relevant property has to be specified and every line of code has to be shown to meet the specifications. In the Flyspeck project, the combinatorial enumeration of tame graphs was verified in this way by Tobias Nipkow and Gertrud Bauer.

There is also a middle ground in which functions are defined algorithmically within the formal system and then executed using an evaluator that is designed for that purpose. There is then a tradeoff between the complexity of the evaluator and the reliability of the result. The verification of the four-color theorem used such a strategy to evaluate the computational component of the proof.

One notable effort along these lines, by Frédéric Chyzak, Assia Mahboubi, Thomas Sibut-Pinote, and Enrico Tassi, yielded a verification of Apéry's celebrated 1973 proof of the irrationality of $\zeta(3)$. The starting point for the project was a Maple worksheet, designed by Bruno Salvy, that carried out the relevant symbolic computation. The group's strategy was to extract algebraic identities from the Maple computations and then construct formal axiomatic proofs of these identities in Coq. A fair amount of work was needed to isolate and manage side conditions that were ignored by Maple, such as showing that a symbolic expression in the denominator of a fraction is nonzero under the ambient hypotheses.

Yet another interesting project was associated with Tucker's solution to Smale's fourteenth problem. To demonstrate the existence of the Lorenz attractor, Tucker enclosed a Poincaré section of the flow defined by the Lorenz equations with small rectangles and showed that each rectangle (together with a cone enclosing the direction in which the attractor is expanding) is mapped by the flow inside another such rectangle (and cone) Tucker, a leading figure in the art of validated computation, relied on careful numeric computation for most of the region, coupled with a detailed analysis of the dynamics around the origin. Quite recently, Fabian Immler was able to verify the numeric computations in Isabelle. To do so, he not only formalized enough of the theory of dynamical systems to express all the relevant claims, but also defined the data structures and representations needed to carry out the computation efficiently and derived enough of their properties to show that the computation meets its specification.

Once again, on the basis of such examples, it may seem bold to predict that formally verified computation will become commonplace in mathematics. The need, however, is pressing. The increasing use of computation to establish mathematical results raises serious concerns as to their correctness, and it is interesting to see how mathematicians struggle to address this. In their 2003 paper, "New Upper Bounds on Sphere Packings. I," Cohn and Elkies provide a brief description of a search algorithm:

> To find a function g [with properties that guarantee an upper bound] . . . , we consider a linear combination of $g_1, g_3, \ldots, g_{4m+3}$, and require it to have a root at 0 and m double roots at $z_1, \ldots,$ z_m. . . . We then choose the locations of $z_1, \ldots, z_m$ to minimize the value r of the last sign change of g. To make this choice, we do a computer search. Specifically, we make an initial guess for the locations of $z_1, \ldots, z_m$, and then see whether we can perturb them to decrease r. We repeat the perturbations until reaching a local optimum.

After presenting the bounds that constitute the main result of the paper, they write:

> These bounds were calculated using a computer. However, the mathematics behind the calculations is rigorous. In particular, we use exact rational arithmetic, and apply Sturm's theorem to count real roots and make sure we do not miss any sign changes.

The passage goes on to explain how they used approximations to real-valued calculations by rational calculations without compromising correctness of the results. In their 2013 paper "The Maximal Number of Exceptional Dehn Surgeries," Lackenby and Meyerhoff turn to the topic of computation:

We now discuss computational issues and responses arising from our parameter space analysis. The computer code was written in C++.

They then proceed to sketch the algorithms they used to carry out the calculations described in the paper, as well as the methods for interval arithmetic, and some of the optimizations they used. They also discuss the use of Snap, a program for studying arithmetic invariants of hyperbolic 3-manifolds, which incorporates exact arithmetic based on algebraic numbers. In their preprint "Universal Quadratic Forms and the 290 Theorem," Bhargava and Hanke are forthright in worrying about the reliability of their computations:

> As with any large computation, the possibility of error is a real issue. This is especially true when using a computer, whose operation can only be viewed intermittently and whose accuracy depends on the reliability of many layers of code beneath the view of all but the most proficient computer scientist. We have taken many steps to ensure the accuracy of our computations, the most important of which are described below.

These steps include checks for correctness, careful management of round-off errors, and, perhaps most importantly, making the source code available on a web page maintained by the authors.

The paper by Cohn and Elkies appeared in the *Annals of Mathematics*, and the one by Lackenby and Meyerhoff appeared in *Inventiones Mathematicae*. The fact that these are two of the most respected journals in mathematics makes it clear that substantial uses of computation have begun to infiltrate the upper echelons of pure mathematics, and the trend is likely to continue. In the passages above, the authors are doing everything they can to address concerns about the reliability of the computations, but the mathematical community does not yet have clear standards for evaluating such results. Are referees expected to read the code and certify the behavior of each subroutine? Are they expected to run the code and, perhaps, subject it to empirical testing? Can they trust the reliability of the software libraries and packages that are invoked? Should authors be required to comment their code sufficiently well for a computer-savvy referee to review it?

Whatever means we develop to address these questions have to scale. Perhaps the bodies of code associated with the examples above are manageable, but what will happen when results rely on code that is even more complicated, and, say, ten times as long? With results like the four-color theorem and Hales' theorem, we are gradually getting past the vain hope that every interesting mathematical theorem will have a humanly surveyable proof. But it seems equally futile to hope that every computational proof will make use of code that can easily be understood, and so the usual difficulties associated with understanding complicated proofs will be paired with similar difficulties in understanding complicated programs.

Formal Search

Formal verification does not have a visceral appeal to most mathematicians: The work can be painstakingly difficult, and the outcome is typically just the confirmation of a result that we had good reason to believe from the start. In that respect, the Pythagorean triple theorem of Heule, Kullmann, and Marek fares much better. Here the outcome of the effort was a new theorem of mathematics, a natural Ramsey-like result, and a very pretty one at that. The result relied on paradigmatic search techniques from the formal methods community, and it seems worthwhile to explore the extent to which such methods can be put to good mathematical work.

To date, such applications of formal methods to mathematics are few and far between. In 1996, William McCune proved the Robbins conjecture, settling a question that had been open since the 1930s as to whether a certain system of equations provided an equivalent axiomatization of Boolean algebras. The result was featured in an article by Gina Kolata in the *New York Times* (1996). But the subject matter was squarely in the field of mathematical logic, and so it is not surprising that an automated theorem prover (in this case, one designed specifically for equational reasoning) could be used for such purposes.

Systems like McCune's can also be used to explore consequences of other first-order axioms. For example, McCune himself showed that the single equation $(w((x^{-1}w)^{-1}z))((yz)^{-1}y) = x$ axiomatizes groups in a language with a binary multiplication and a unary inverse, and Kenneth Kunen later showed that this is the shortest such axiom. Kunen went on to use interactive theorem provers to contribute notable results to the theory of nonassociative structures such as loops and quasigroups. (More examples of this sort are discussed in Beeson [2004].)

Since the beginning of this century, propositional satisfiability solvers have been the killer app for formal methods, permitting algorithmic solutions to problems that were previously out of reach. On the heels of the Pythagorean triples problem, Heule has recently established that the Schur number $S(5)$ is equal to 160; in other words, there is a five-coloring of the integers from 1 to 160 with no monochromatic triple a, b, c with $a + b = c$, but no such coloring of the integers from 1 to 161.

A SAT solver had a role to play in work on the Erdős discrepancy problem. Consider a sequence $(x_i)_{i>0}$, where each x_i is ± 1, and consider sums of this sequence along multiples of a fixed positive integer, such as $x_1 + x_2 + x_3 + \ldots$ and $x_2 + x_4 + x_6 + \ldots$ and $x_3 + x_6 + x_9 + \ldots$. In the 1930s, Erdős asked whether it is possible to keep the absolute value of such sums—representing the discrepancy between the number of +1s and −1s along the sequence—uniformly bounded. In other words, he asked whether there are a sequence (x_i) and a value C such that for every n and d, $\left| \sum_{i=1}^{n} x_{id} \right| \leq C$, and he conjectured that no such pair exists. In 2010, Tim Gowers launched the collaborative Polymath5 project on his blog to work on the problem. In 2014, Boris Konev and Alexei Lisitsa used a SAT solver to provide a partial result, namely, that there is no sequence satisfying the conclusion with $C = 2$. Specifically, they showed that there is a finite sequence $x_1, \ldots, x_{1,160}$, with discrepancy at most 2, but no such sequence of length 1,161. The following year, Terence Tao proved the full conjecture, with a conventional proof. This proof was a much more striking achievement, but we still have Konev and Lisitsa, and a SAT solver, to thank for exact bounds in the case $C = 2$. SAT solvers have been applied to other combinatorial problems as well.

The line between discovery and verification is not sharp. Anyone writing a search procedure does so with the intention that the results it produces will be reliable, but, as with any piece of software, as the code becomes more complex, it becomes increasingly necessary to have mechanisms to ensure that the results are correct. This requirement is especially important with the use of powerful search tools, which rely on complicated tricks and heuristics to improve performance at the risk of compromising soundness. It is important that the solution to the Pythagorean triples problem produced a formal proof that could be verified independently, and, in fact, that proof has been checked by three proof checkers which themselves have been formally verified, one in Isabelle, one in Coq, and one in a theorem prover named ACL2. This confirmation provides a high degree of confidence in the correctness of the result.

Today, the use of formal methods in discovery is even less advanced than the use of formal methods in verification. The results described here depend, for the most part, on finding consequences of first-order axioms for algebraic structures, searching for finite objects satisfying combinatorial constraints, or ruling out the existence of such objects by exhaustive enumeration. It is not surprising that computers can be used to exhaust a large number of finite cases, but few mathematical problems are presented to us in that form. And spinning out consequences of algebraic axioms is a far cry from discovering consequences of rich mathematical assumptions involving heterogeneous structures and mappings between them.

But just as pure mathematicians have discovered uses for computation in number theory, algebraic topology, differential geometry, and discrete geometry, one would expect to find similarly diverse applications for formal search methods. The problem may simply be that researchers in these fields do not yet have a sense of what formal search methods can do, whereas the computer scientists who develop them do not have the expertise needed to identify the mathematical domains of application. If that is the case, it is only a matter of getting the communities to work more closely together. Combinatorics is a natural place to start because the core concepts are easily accessible and familiar to computer scientists. But it will take real mathematical effort to understand how problems in other domains can be reduced to the task of finding finite pieces of data or ruling out the existence of such data by considering sufficiently many cases.

Indeed, for all we know, there may be lots of lovely theorems of mathematics that can *only* be proved that way. For the last 2,000 years, we have been looking for proofs of a certain kind because those are the proofs that we can survey and understand. In that respect, we may be like the drunkard looking for his keys under a streetlamp even though he lost them a block away, because that is where the light is. We should be open to the possibility that new technologies can open new mathematical vistas and afford new types of mathematical understanding.

The prospect of ceding a substantial role in mathematical reasoning to the computer may be disconcerting, but it should also be exhilarating, and we should look forward to seeing where the technology takes us.

Digital Infrastructure

Contemporary digital technologies for storage, search, and communication of information provide another market for formal methods in mathematics. Mathematicians now routinely download papers, search the web for mathematical results, post questions on Math Overflow, typeset papers using LaTeX, and exchange mathematical content via e-mail. Digital representations of mathematical knowledge are therefore central to the mathematical process. It stands to reason that mathematics can benefit from having better representations and better tools to manage them.

TeX and LaTeX have transformed mathematical dissemination and communication by providing precise means for specifying the appearance of mathematical expressions. MathML, building on XML, goes a step further, providing markup to specify the *meaning* of mathematical expressions as well. But MathML stops short of providing a foundational specification language, which is clearly desirable: Imagine being able to find the statement of a theorem online, and then being able to look up the meaning of each defined term, all the way down to the primitives of an axiomatic system if necessary. That capability would provide clarity and uniformity and would help ensure that the results we find mean what we think they mean. The availability of such formal specifications would also support verification: We could have a shared public record of which results have been mechanically verified and how, and we could use theorems from a public repository to verify our own local results. Automated reasoning tools could make use of such background knowledge, and could, in turn, be used to support a more robust search. Contemporary *sledgehammer* tools for interactive theorem provers rely on heuristics to extract relevant theorems from a database and then use them to carry out a given inference. With such technology, one could ask whether a given statement is equivalent to, or an easy consequence of, something in a shared repository of known facts.

For all these purposes, formal specifications are essential. As a first step toward obtaining them, Hales has recently launched a Formal

Abstracts project, which is designed to encourage mathematicians to write formal abstracts of their papers. To process and check the definitions, he has chosen an interactive theorem prover called Lean, an open source project led by Leonardo de Moura at Microsoft Research (to which I am a contributor). In the coming years, the Formal Abstracts project plans to seed the repository with core definitions from all branches of mathematics and to develop guidelines, tools, and infrastructure to support widespread use.

Conclusions

In the summer of 2017, the Isaac Newton Institute hosted a six-week workshop, called Big Proof, dedicated to the technologies described here.[3] As part of a panel discussion, Timothy Gowers gave a frank assessment of the new technology and the potential interest to mathematicians. He observed that the phrase "interactive proof assistant" is rather appealing until one learns that such assistants actually make proving a theorem a lot more difficult. The fact that a substantial body of undergraduate mathematics has been formalized is generally unexciting to the working mathematician, and existing tools currently offer little to improve our mathematical lives.

Gowers did enumerate three technologies that he felt would have widespread appeal. The first is a bona fide proof assistant that could work out small lemmas and results, at the level of a capable graduate student. The second is genuine search technology that can tell us whether a given fact is currently known, either because we would like to use it in a proof or because we think we have a proof and are wondering whether it is worthwhile to work out the details. The third is a real proof checker, that is, something we can call when we think we have proved something and want confirmation that we have not made a mistake.

We are not there yet, but such technology seems to be within reach. There are no apparent conceptual hurdles that need to be overcome, though getting to that point will require a good deal of careful thought, clever engineering, experimentation, and hard work. And even before tools like these are ready for everyday use, we can hope to find pockets of mathematics where the methods provide a clear advantage:

proofs that rely on nontrivial calculations, subtle arguments for which a proof assistant can provide significant validation, and problems that are more easily amenable to search techniques. Verification is not an all-or-nothing affair. Short of a fully formalized axiomatic proof, formalizing a particularly knotty or subtle lemma or verifying a key computation can lend confidence to the correctness of a result. Even just formalizing definitions and the statements of key theorems, as proposed by the Formal Abstracts project, adds helpful clarity and precision. Formal methods can also be used in education: If we teach students how to write formal proofs and informal proofs at the same time, the two perspectives reinforce one another (Avigad et al. 2017).

The mathematics community needs to put some skin in the game, however. Proving theorems is not like verifying software, and computer scientists do not earn promotions or secure funding by making mathematicians happy. We need to buy into the technology if we want to reap the benefits.

To that end, institutional inertia needs to be overcome. Senior mathematicians generally do not have time to invest in developing a new technology, and it is hard enough to learn how to use the new tools, let alone contribute to their improvement. The younger generation of mathematicians has prodigious energy and computer savvy, but younger researchers would be ill-advised to invest time and effort in formal methods if it will only set back their careers. To allow them to explore the new methods, we need to give them credit for publications in journals and conferences in computer science, and recognize that the mathematical benefits will come only gradually. Ultimately, if we want to see useful technologies for mathematics, we need to hire mathematicians to develop them.

The history of mathematics is a history of doing whatever it takes to extend our cognitive reach and of designing concepts and methods that augment our capacities to understand. The computer is nothing more than a tool in that respect, but it is one that fundamentally expands the range of structures we can discover and the kinds of truths we can reliably come to know. This is as exciting a time as any in the history of mathematics, and even though we can only speculate as to what the future will bring, it should be clear that the technologies before us are well worth exploring.

Acknowledgment

I am grateful to Jasmin Blanchette and Robert Y. Lewis for corrections, suggestions, and improvements.

Notes

1. Hales provided an engaging account of the refereeing process and the motivation behind the Flyspeck project in a talk presented to the Isaac Newton Institute in the summer of 2017, www.newton.ac.uk/seminar/20170710100011001.

2. www.math.ias.edu/~vladimir/Site3/Univalent_Foundations_ files/2014_IAS.pdf.

3. Talks delivered at the program are available online at www. newton.ac.uk/event/bpr.

References

Jeremy Avigad and John Harrison, Formally verified mathematics, *Commun. ACM* 57(4):66–75, 2014.

Jeremy Avigad, Robert Y. Lewis, and Floris van Doorn, *Logic and Proof*, online textbook, 2017, https://github.com/leanprover/logic_and_proof.

Michael J. Beeson, The mechanization of mathematics. In *Alan Turing: Life and Legacy of a Great Thinker*, Springer, Berlin, 2004, pp. 77–134. MR2172456.

Henry Cohn and Noam Elkies, New upper bounds on sphere packings. I, *Annals of Mathematics* 157 (2003), 689–714.

Kurt Gödel, Über formal unentscheidbare Sätze der Principia Mathematica und verwandter Systeme I. *Monatsh. Math. Phys.*, 38(1):173–198, 1931. Reprinted with English translation in *Kurt Gödel: Collected Works*, vol. 1, Feferman et al, eds., Oxford University Press, New York, 1986, pp. 144–195. MR1549910.

Thomas C. Hales, Mathematics in the age of the Turing machine. In *Turing's Legacy: Developments from Turing's Ideas in Logic*, vol. 42 of Lecture Notes in Logic, pp. 253–298. Assoc. Symbol. Logic, La Jolla, CA, 2014. MR3497663.

Gina Kolata, Computer math proof shows reasoning power. *New York Times*, December 10, 1996.

Marc Lackenby and Robert Meyerhoff, The maximal number of exceptional Dehn surgeries. *Inventiones Mathematicae* 191 (2013), 341–382.

Melvyn B. Nathanson, Desperately seeking mathematical proof, *Math. Intelligencer*, 31 (2):8–10, 2009. MR2505014.

Mathematics as an Empirical Phenomenon, Subject to Modeling

Reuben Hersh

Modeling is a central feature of contemporary empirical science. There is mathematical modeling, there is computer modeling, and there is statistical modeling, which is halfway between. We may recall older models: plaster models of mathematical surfaces, stick-and-ball models of molecules, and the model airplanes that used to be so popular but now have been promoted into drones.

Today the scholarly or scientific study of any phenomenon, whether physical, biological, or social, implicitly or explicitly uses a model of that phenomenon. A physicist studying heat conduction, for example, may model heat conduction as a fluid flow, or as propagation of kinetic energy of molecules, or as a relativistic or quantum mechanical action. Different models serve different purposes. Setting up a model involves focusing on features of the phenomenon that are compatible with the methodology being proposed and neglecting features that are not compatible with it. A mathematical model in applied science explicitly refrains from attempting to be a complete picture of the phenomenon being modeled.

Mathematical modeling is the modern version of both applied mathematics and theoretical physics. In earlier times, one proposed not a model but a theory. By talking today of a model rather than a theory, one acknowledges that the way one studies the phenomenon is not unique; it could also be studied other ways. One's model need not claim to be unique or final. It merits consideration if it provides an insight that is not better provided by some other model.

It is disorienting to think of mathematics as the thing being modeled because much of mathematics, starting with elementary arithmetic, already *is* a model of a physical action. Arithmetic, for instance, models the human action of counting.

The philosophy of mathematics, when studying the "positions" of formalism, constructivism, platonism, and so on, is studying models of mathematics, which is in large part a model. It studies second-order models! (Other critical fields like literary and art criticism are also studying models of models.) Being a study of second-order models, the philosophy of mathematics constitutes still a higher order of modeling—a third-order model!

In this article, I make a few suggestions about the modeling of mathematics.

Empirical Studies of Mathematics

To study any phenomenon, a scholar or scientist must conceptualize it in one way or another. She must focus on some aspects and leave others aside. That is to say, she models it.

Mathematical knowledge and mathematical activity are observable phenomena, already present in the world, already out there, before philosophers, logicians, neuroscientists, or behavioral scientists proceed to study them.

The empirical modeling of social phenomena is a whole industry. Mathematical models, statistical models, and computer models strive to squeeze some understanding out of the big data that are swamping everyone. Mathematical *activity* (in contrast to mathematical *content*) is one of these social phenomena. It is modeled by neuroscience, by logic, by the history of mathematics, by the psychology of mathematics, by anthropology, and by sociology. These models must use verbal modeling for phenomena that are not quantifiable—the familiar psychological and interpersonal variables of daily life, including mathematical life.

Recognizing mathematical behavior and mathematical life as empirical phenomena, we would expect to use various different models, each focusing on a particular aspect of mathematical behavior. Some of these models might be mathematical. For such models, there would be a certain reflexivity or self-reference, since the model then would be part of the phenomenon being modeled.

History, logic, neuroscience, psychology, and other sciences offer different models of mathematics, each focusing on the aspects that are accessible to its method of investigation. Different studies of mathematical life overlap; they have interconnections, but still, each works to its own special standards and criteria. Historians are historians first of all and likewise educators, neuroscientists, and so on. Each special field studying math has its own model of mathematics.

Each of these fields has its particular definition of mathematics. Rival definitions could provoke disagreement, even conflict. Disagreement and conflict are sometimes fruitful or instructive, but often they are unproductive and futile. I hope to convince some members of each profession that his or her viewpoint is not the only one that is permissible. I try to do justice to all, despite the bias from a lifetime as a mathematician.

Let us look separately at four of the math-studying disciplines and their models.

Logic

Among existing models of mathematics, the giant branch of applied logic called *formalized mathematics* is by far the most prestigious and successful. Being at once a model of mathematics and a branch of mathematics, it has a fascinating self-reflexivity. Its famous achievements are at the height of mathematical depth. Proudly and justifiably, it excludes the psychological, the historical, the personal, the contingent, and the transitory aspects of mathematics.

Related but distinct is the recent modeling of mathematical proofs in actual code that runs on an actual machine. Such programs come close to guaranteeing that a proof is complete and correct.

Logic sees mathematics as a collection of virtual inscriptions— declarative sentences that could in principle be written down. On the basis of that vision, it offers a model: formal deductions from formal axioms to formal conclusions—formalized mathematics. This vision itself is mathematical. Mathematical logic is a branch of mathematics, and whatever it is saying about mathematics, it is saying about itself— self-reference. Its best results are among the most beautiful in all of mathematics (Godel's incompleteness theorems, Robinson's nonstandard analysis).

This powerful model makes no attempt to resemble what real mathematicians really do. That project is left to others. The logician's view of mathematics can be briefly stated (perhaps oversimplified) as "a branch of applied logic."

The competition between category theory and set theory, for the position of "foundation," can be regarded as a competition within logic, for two alternative logical foundations. Ordinary working mathematicians see them as two alternative models, either of which one may choose, as seems best for any purpose.

NEUROSCIENCE

The work of neuroscientists like Dehaene (1997) is a beginning on the fascinating project of finding how and where mathematical activity takes place on the biophysical level of flesh and blood. Neuroscience models mathematics as an activity of the nervous system. It looks at electrochemical processes in the nervous system of the mathematician. There it seeks to find correlates of her mathematical process. Localization in the brain will become increasingly accurate as new research technologies are invented. With accurate localization, it may become possible to observe activity in specific brain processes synchronized with conscious mathematical thought. Already, Changeux, in Connes and Changeux (1995) argues forcefully that mathematics is nothing but a brain process.

The neuroscientist's model of mathematics can be summarized (a bit oversimplified) as "a certain kind of activity of the brain, the sense organs, and sensory nerves."

THE HISTORY OF MATHEMATICS

The history of mathematics is studied by mathematicians as well as historians. History models mathematics as a segment of the ongoing story of human culture. Mathematicians are likely to see the past through the eyes of the present and ask, "Was it important? natural? deep? surprising? elegant?" The historian sees mathematics as a thread in the ever-growing web of human life, intimately interwoven with finance and technology, with war and peace. Today's mathematics is the culmination of all that has happened before now, yet to future viewpoints it will seem like a brief, outmoded stage of the past.

Philosophy

Many philosophers have proposed models of mathematics, but without explicitly situating their work in the context of modeling. Lakatos' *Proofs and Refutations* (1976) presents a classroom drama about the Descartes-Euler formula. The problem is to find the correct definition of *polyhedron* to make the Descartes-Euler formula applicable. The successive refinement by examples and counterexamples is implicitly being suggested as a model for mathematical research in general. Of course, critics of Lakatos found defects in this model. His neat reconstruction overlooked or omitted inconvenient historical facts. Lakatos argued that his rational reconstruction was more instructive than history itself! This is amusing or outrageous, depending on how seriously you take these matters. It is a clear example of violating the zeroth law of modeling, which is this: Never confuse or identify the model with the phenomenon!

Philip Kitcher's *The Nature of Mathematical Knowledge* (1983) sought to explain how mathematics grows, how new mathematical entities are created. He gave five distinct driving forces to account for this. Feferman (1998), in constructing the smallest system of logic that is big enough to support classical mathematics, is also offering us a model of mathematics. Grosholz (2007) in focusing on what she calls *ampliative* moves in mathematical research, is modeling mathematical activity. Cellucci (2006), in arguing that plausible reasoning rather than deductive reasoning is the essential mathematical activity, is also proposing a model of mathematics. In *A Subject with No Object*, Burgess and Rosen (1997) conclude that nominalist reconstructions of mathematics help us better understand mathematics—even though nominalism (they argue) is not tenable as a philosophical position. This short list reflects my own reading and interests. Many others could be mentioned.

Analogous to the well-established interaction of the history of science and the philosophy of science, there has been some fruitful interaction between the philosophy of mathematics and the history of mathematics. One disappointing example was the great French number theorist André Weil (1978), who in his later years took an interest in history and declared that no two fields have less in common than the philosophy of math and the history of math. The philosopher-historian Imre Lakatos, on the other hand, wrote that without philosophy, history is lame, and without history, philosophy is blind. Or maybe it is the other way around. Each model is important; none should be ignored.

The collaboration between philosopher Mark Johnson and linguist George Lakoff is exemplary. (*Where Mathematics Comes From* (2000) by Lakoff and Rafael Nuñez, is a major contribution to our understanding of the nature of mathematics.)

There are some eccentric, philosophically oriented mathematicians. We try to untangle our own and each other's actions and contributions. We do not always manage to separate the content of mathematics from the activity of mathematics, for to us they are inseparable. We are not offering contributions to philosophy. We are not philosophers, as some philosophers politely inform us. We merely try to report faithfully and accurately what we really do. We are kindly tolerated by our fellow mathematicians and are considered gadflies by the dominant philosophers.

Byers (2010) introduced ambiguity as an essential aspect of mathematics and a driving force that leads to the creation of new mathematics.

Several leading mathematicians have written accounts of their own experience in a phenomenological vein; I quote them in "How Mathematicians Convince Each Other," one of the chapters in *Experiencing Mathematics* (Hersh 2014).

My own recent account of mathematicians' proofs (Hersh 2014) is another possible model of mathematics. Here it is: A mathematician possesses a mental model of the mathematical entity she works on. This internal mental model is accessible to her direct observation and manipulation. At the same time, it is socially and culturally controlled to conform to the mathematics community's collective model of the entity in question. The mathematician observes a property of her own internal model of that mathematical entity. Then she must find a recipe, a set of instructions that enables other competent, qualified mathematicians to observe the corresponding property of their corresponding mental models. That recipe is the proof. It establishes that property of the mathematical entity. This is a verbal, descriptive model. Like any model, it focuses on certain specific features of the situation and, by attending to those features, seeks to explain what is going on.

The discussion up to this point has left out the far greater part of ongoing mathematical activity—that is, schooling. Teaching and learning. Education.

Teachers and educators will be included in any future comprehensive science of mathematics. They observe a lot and have a lot to say about it. Paul Ernest, in his book *Social Constructivism in the Philosophy of Mathematics* (1997) follows Lakatos (1976) and Wittgenstein in building his social constructivist model.

Mathematics education has urgent questions to answer. What should be the goals of math education? What methods could be more effective than the present disastrously flawed ones? Mathematics educators carry on research to answer these questions. Their efforts would be greatly facilitated by a well-established overall study of the nature of mathematics.

Why not seek for a unified, distinct scholarly activity of *mathematics studies*: the study of mathematical activity and behavior? Mathematics studies could be established and recognized, in a way comparable to the way that linguistics has established itself, as the study of mathematical behavior, by all possible methods. Institutionally, it would not interfere with or compete with mathematics departments, any more than linguistics departments impinge on or interfere with the long-established departments of English literature, French literature, Russian literature, and so on.

Rather than disdain the aspect of mathematics as an ongoing activity of actual people, philosophers could seek to deepen and unify it. How do different models fit together? How do they fail to fit together? What are their contributions and their shortcomings? What is still missing? This role for the philosophy of mathematics would be higher than the one usually assigned to it.

A coherent inclusive study of the nature of mathematics would contribute to our understanding of problem solving in general. Solving problems is how progress is made in all of science and technology. The synthesizing energy to achieve such a result would be a worthy and inspiring task for philosophy.

About Modeling and the Philosophy of Mathematics

Turning now to the content of mathematics rather than the activity, we are in the realm of present-day philosophy of mathematics.

Philosophers of mathematics seem to be classified by their "positions," as though philosophy of mathematics consisted mainly of choosing a position and then arguing against other positions. I take Stewart Shapiro's *The Oxford Handbook of Philosophy of Mathematics and Logic* (2005) as a respected representative. "I now present sketches of some main positions in the philosophy of mathematics," he writes.

Six positions appear in the table of contents, and five of them get two chapters, pro and con. Between chapters expounding logicism, intuitionism, naturalism, nominalism, and structuralism are chapters reconsidering structuralism, nominalism, naturalism, intuitionism, and logicism. "One of these chapters is sympathetic to at least one variation on the view in question, and the other 'reconsiders.'," Shapiro says in the introduction. Formalism gets only one chapter; evidently it does not need to be reconsidered.

"A survey of the recent literature shows that there is no consensus on the logical connections between the two realist theses or their negations. Each of the four possible positions is articulated and defended by established philosophers of mathematics."

"Taking a position" on the nature of mathematics looks very much like the vice of *essentialism*—claiming that some description of a phenomenon captures what that phenomenon "really is" and then trying to force observations of that phenomenon to fit into that claimed essence. Rival essentialisms can argue for a long time; there is no way either can force the other to capitulate.

Such is the story of mathematical platonism and mathematical antiplatonism. Balaguer (2001, 2013) has even published two books proving that neither of those two positions can *ever* be proved or disproved. "He concludes by arguing that it is not simply that we do not currently have any good arguments for or against platonism but that we could never have such an argument." Balaguer's conclusion is correct. It is impossible in principle to *prove or disprove* any model of any phenomenon, for the phenomenon itself is prior to, independent of, our formalization, and cannot be regarded as or reduced to a term in a formal argument.

One natural model for mathematics is as story or narrative. Thomas (2007) suggests such a model. Thinking of mathematical proofs or theories as stories has both obvious merits and defects. Pursuing its merits might have payoffs in research or in teaching. That would be different from being a fictionalist—taking the position that mathematics *is* fiction. Thomas (2014) has also suggested litigation and playing a game as models for mathematical activity.

Another natural model for mathematics is as a structure of structures (whether *ante rem* or otherwise). It is easy to see the merits of such a model and not hard to think of some defects. Pursuing the merits might have a payoff, in benefiting research or teaching. This pursuit would be a different matter from being a structuralist—taking the position that mathematics *is* structure.

The model of mathematics as a formal axiomatic structure is an immense success, settling Hilbert's first and tenth problems and providing tools for mathematics like nonstandard analysis. It is a branch of mathematics while simultaneously being a model of mathematics, so it possesses a fascinating and bewildering reflexivity. Enjoying these benefits does not require one to be a formalist—to claim that mathematics *is* an axiomatic structure in a formal language. Thurston (2006) testifies to the needless confusion and disorientation which that formalist claim causes to beginners in mathematical research.

If a philosopher of mathematics regarded his preferred "position" as a model rather than a theory, he might coexist and interact more easily. Structuralism, intuitionism, naturalism, nominalism/fictionalism, and realism/platonism each has strengths and weaknesses as a model for mathematics. Perhaps the most natural and appealing philosophical tendency for modeling mathematics is phenomenology. The phenomenological investigations of Merleau-Ponty looked at *outer* perception, especially vision. A phenomenological approach to mathematical behavior would try to capture an *inner* perception, the mathematician's encounter with her own mathematical entity.

If we looked at these theories as models rather than as theories, it would hardly be necessary to argue that each one falls short of capturing all the major properties of mathematics, for no model of any empirical phenomenon can claim to do that. The test for models is whether they are useful or illuminating, not whether they are complete or final.

Different models are both competitive and complementary. Their standing depends on their benefits in practice. If the philosophy of mathematics were seen as modeling rather than as taking positions, it might consider paying attention to mathematics research and mathematics teaching as testing grounds for its models.

Can we imagine these rival schools settling for the status of alternative models, each dealing with its own part of the phenomenon of interest, each aspiring to offer some insight and understanding? The structuralist, platonist, and nominalist could accept that in the content of mathematics, even more than in heat conduction or electric currents, no single model is complete. Progress would be facilitated by encouraging each in his own contribution, noticing how different models overlap and connect, and proposing when a new model may be needed. A modeling paradigm would substitute competition for conflict. One philosophical modeler would allow the other modeler his or her model. By their fruits would they be judged.

Frege expelled psychologism and historicism from respectable philosophy of mathematics. Nevertheless, it is undeniable that mathematics is a historical entity and that mathematical work and activity are mental work and activity. Its history and its psychology are essential features of mathematics. We cannot hope to understand mathematical activity while forbidding attention to the mathematician's mind.

As ideologies, historicism and psychologism are one-sided and incomplete, as was logicism's reduction of mathematics to logic. We value and admire logic without succumbing to logicism. We can see the need for the history of mathematics and the psychology of mathematics without committing ourselves to historicism or psychologism.

The argument among fictionalists, platonists, and structuralists seems to suppose that some such theory could be or should be the actual truth. But mathematics is too complex, varied, and elaborate to be encompassed in any model. An all-inclusive model would be like the map in the famous story by Borges—perfect and inclusive because it was identical to the territory it was mapping.

Formalists, logicists, constructivists, and so on can each try to provide understanding without discrediting each other, just as the continuum model of fluids does not contradict or interfere with the kinetic model.

Some Elementary Number Theory

Since nothing could be more tedious than 20 pages of theorizing about mathematics without a drop of actual mathematics, I end with an example from the student magazine *Eureka* (Hersh 2013), which was republished with a different title in the *College Mathematics Journal* (Hersh 2012). It is an amusing, instructive little sample of mathematicians' proofs and a possible test case for different models of mathematics.

A high school exercise is to find a formula for the sum of the first n cubes. You quickly sum

$$1 + 8 + 27 + 64 + 125 \ldots$$

and find the successive sums

$$1, 9, 36, 100, 225 \ldots$$

You immediately notice that these are the squares of

$$1, 3, 6, 10, 15$$

which are the sums of the first n integers for

$$n = 1, 2, 3, 4, \text{ and } 5.$$

If we denote the sum of the pth powers of the integers, from the first up to the nth, as the polynomial $S_p(n)$, which always has degree $p + 1$, then our discovery about the sum of cubes is compact:

$$S_3(n) = [S_1(n)]^2$$

What is the reason for this surprising relationship? Is it just a coincidence?

A simple trick explains the mystery. We see that the sums of odd powers—the first, third, fifth, or seventh powers, and so on—are always polynomials in the sum of the first n integers. If you like, you could call this a "theorem."

I will give you instructions. To start, just make a table of the sums of pth powers of the integers, with

$p = 0$ in the first row,
$p = 1$ in the second row,
$p = 2$ in the third row, and
$p = 3$ in the fourth row.

Instead of starting each row at the left side of the page, start in the middle of the page, like this:

0	1	2	2	4	5 ...
0	1	3	6	10	15 ...
0	1	5	14	30	55 ...
0	1	9	36	100	225 ...

Now notice that nothing prevents you from extending these rows to the left—by successive *subtractions* of powers of integers, instead of adding!

In the odd rows, subtracting negative values, you obtain positive entries. Here is what you get:

−5	−4	−3	−2	−1	0	0	1	2	3	4	5
15	10	6	3	1	0	0	1	3	6	10	15
−55	−30	−14	−5	−1	0	0	1	5	14	30	55
225	100	36	9	1	0	0	1	9	36	100	225

The double appearance of 0 in each row results from the fact that in the successive subtractions, a subtraction of 0 occurs between the subtractions of 1 to the pth power and (-1) to the pth power.

Notice the symmetry between the right and left half of each row. The symmetry of the first and third row is opposite to the symmetry of the second and fourth. These two opposite kinds of symmetry are called "odd" and "even," respectively.

(That is because the graphs of the odd and even power functions have those two opposite kinds of symmetry. The even powers 2, 4, and so on, have the same values in the negative direction as in the positive direction. For degree 2, the graph is the familiar parabola of $y = x^2$, with axis of symmetry on the y-axis. The fourth power, sixth power, and so on have more complicated graphs, but they all are symmetric with respect to the vertical axis. The graphs of the odd powers, on the other hand (the first, third, fifth, and so on), are symmetric in the opposite way, taking negative values in the negative direction (in the "third quadrant") and symmetric with respect to a point, the origin of coordinates.)

The two opposite symmetries in your little table suggest that the sum functions of the integers raised to even powers are odd polynomials, and the sums of odd powers are even polynomials.

Continuing to the left is done by *subtracting* $(-n)^p$. For the odd powers p, this is *negative*, so the result is *adding* n^p. That is the same as what you would do to continue *to the right*, adding the pth power of the next integer. Therefore, the observed symmetry for odd powers continues for all n, and for every odd p, not just the $p = 1$ and $p = 3$ that we can read off our little table.

But surprise! The center of symmetry is not at

$$n = 0$$

but halfway between 0 and −1! Therefore, as the table shows, for odd p the polynomial $S_p(n)$ satisfies the shifted symmetry identity

$$S_p(-n) = S_p(n - 1)$$

Therefore, for odd p, the squares, fourth powers, and higher terms of $S_p(n)$ are even powers of $(n + 1/2)$. A sum of those *even* powers is the same thing as a sum of *all* powers of $(n + 1/2)^2$, which would be called "a polynomial in $(n + 1/2)^2$." To complete our proof, we need only show that

$$(n + 1/2)^2 = 2S_1 + 1/4$$

Now $S_1(n)$ is familiar; everybody knows that it is equal to

$$n(n + 1)/2$$

(There is a much-repeated anecdote about how this was discovered by the famous Gauss when he was a little boy in school.)

So then, multiplying out,

$$2S_1 = n^2 + n$$

We do a little high school algebra:

$$(n + 1/2)^2 = n^2 + n + 1/4 = 2S_1 + 1/4$$

so for odd p we do have S_p as a polynomial in S_1, as claimed.

I leave it to any energetic reader to work out $S_5(n)$ as a polynomial in $S_1(n)$. Since S_5 has degree 6, and S_1 is quadratic, S_5 will be cubic as a polynomial in S_1. There are only three coefficients to be calculated!

This little proof in elementary number theory never even needed to state an axiom or hypothesis. The rules of arithmetic and polynomial algebra did not need to be made explicit, any more than the rules of first-order logic. Without an axiom or a hypothesis or a premise, where was the logic?

Given an interesting question, we dove right into the mathematics and swam through it to reach the answer. We started out, you and I, each possessing our own internal model of mathematical tables, of the integers, and of polynomials in one variable. These models match; they are congruent. In particular, we agree that an odd power of a negative number is negative and that subtracting a negative number results in adding a positive number.

I noticed that continuing the table to the left led to interesting insights. So I gave you instructions that would lead you to those insights. You followed them and became convinced. My list of instructions is the proof!

One could elaborate this example into formalized logic. But what for? More useful would be making it a test for competing models of mathematics (formerly "positions"). How would the structuralist account for it? The nominalist, the constructivist, the platonist, the intuitionist? Which account is more illuminating? Which is more credible? How do they fit together? Are any of them incompatible with each other?

You may wonder, "Am I serious, asking a philosopher to take up modeling, instead of arguing for his chosen position against opposing positions?"

Yes. I am serious. The philosopher will then be more ready to collaborate with historians and cognitive scientists. The prospect for an integrated field of mathematics studies will improve.

However, such a turn is not likely to be made by many. If philosophy is all about "taking a position" and arguing against other positions, a switch from position taking to modeling might bring a loss of standing among philosophers.

Acknowledgments

I value the contributions to the understanding of mathematics made by Carlo Cellucci, Emily Grosholz, George Lakoff and Rafael Nuñez, David Ruelle, Paul Livingston, Philip Kitcher, Paul Ernest, Mark Steiner, William Byers, Mary Tiles, Fernando Zalamea, and Penelope Maddy. I thank Vera John-Steiner, Stephen Pollard, Carlo Cellucci, and Robert Thomas for their suggestions for improving this article.

References

Balaguer, M. (2001). *Platonism and anti-platonism in mathematics.* Oxford, U.K.: Oxford University Press.

———— (2013). *A guide for the perplexed: What mathematicians need to know to understand philosophers of mathematics.* http://sigmaa.maa.org/pom/PomSigmaa/Balaguer1-13.pdf.

Burgesss, J. P., and Rosen, G. (1997). *A subject with no object.* Oxford, U.K.: Oxford University Press.

Byers, W. (2010). *How mathematicians think.* Princeton, NJ: Princeton University Press.

Cellucci, C. (2006). Introduction to *Filosofia e matematica.* In R. Hersh (Ed.), *18 Unconventional essays on the nature of mathematics* (pp. 17–36). Berlin: Springer.

Connes, A., and Changeux, J.-P. (1995). *Conversations on mind, matter and mathematics.* Princeton, NJ: Princeton University Press.

Dehaene, S. (1997). *The number sense.* Oxford, U.K.: Oxford University Press.

Ernest, P. (1997). *Social constructivism in the philosophy of mathematics.* Albany, NY: SUNY Press.

Feferman, S. (1998). *In the light of logic.* Oxford, U.K.: Oxford University Press.

Grosholz, E. (2007). *Representation and productive ambiguity in mathematics and the sciences.* Oxford, U.K.: Oxford University Press.

Hersh, R. (2012). Why the Faulhaber polynomials are sums of even or odd powers of (*n* + 1/2). *College Mathematics Journal, 43*(4), 322–324.

———— (2013). On mathematical method and mathematical proof, with an example from elementary algebra. *Eureka, 63,* 7–8.

———— (2014). *Experiencing mathematics.* Providence, RI: American Mathematical Society.

Kitcher, P. (1983). *The nature of mathematical knowledge.* Oxford, U.K.: Oxford University Press.

Lakatos, I. (1976). *Proofs and refutations.* Cambridge, U.K.: Cambridge University Press.

Lakoff, G., and Nuñez, R. (2000). *Where mathematics comes from.* New York: Basic Books.

Ruelle, D. (2007). *The mathematician's brain.* Princeton, NJ: Princeton University Press.

Shapiro, S. (2005). *The Oxford handbook of philosophy of mathematics and logic.* Oxford, U.K.: Oxford University Press.

Thomas, R. (2007). The comparison of mathematics with narrative. In B. van Kerkhove and J. P. van Bendegem (Eds.), *Perspectives on mathematical practices* (pp. 43–60). Berlin: Springer.

———— (2014). *The judicial analogy for mathematical publication.* Paper delivered at the meeting of the Canadian Society for History and Philosophy of Mathematics, May 25, Brock University, St. Catharines, Ontario, Canada.

Thurston, W. (2006). On proof and progress in mathematics. In R. Hersh (Ed.), *18 Unconventional essays on the nature of mathematics* (pp. 37–55). Berlin: Springer.

Tiles, M. (1991). *Mathematics and the image of reason.* London: Routledge.

Weil, A. (1978). History of mathematics: Why and how. In *Proceedings of the International congress of mathematicians,* Helsinki, Finland.

Does 2 + 3 = 5?
In Defense of a Near Absurdity

MARY LENG

In the last issue of this magazine, James Robert Brown (Brown 2017) asked, "Is anyone really agnostic about 2 + 3 = 5, and willing only to give assent to PA → 2 + 3 = 5?" (where PA stands for the Peano axioms for arithmetic). In fact, Brown should qualify his "anyone" with "anyone not already hopelessly corrupted by philosophy," since, as he knows full well, there are plenty of so-called *nominalist* philosophers—myself included—who, wishing to avoid commitment to abstract (that is, nonspatiotemporal, acausal, mind- and language-independent) objects, take precisely this attitude to mathematical claims.

Why on Earth might one question such a basic claim as "2 + 3 = 5"? First of all, we should be clear about what is not being questioned. That two apples plus three more apples makes five apples is not something that is in question, and neither is the generalized version of this, "For any *F*, if there are two *F*s, and three more *F*s, then there are five *F*s." Of course, this generalization may fail for some *F*s (think of rabbits or raindrops), but suitably qualified so that we only plug in the right kind of predicates as replacements for *F*, this generalization does not worry nominalist philosophers of mathematics—indeed, each of its instances is a straightforward logical truth expressible and derivable in first-order predicate logic, without any mention of numbers at all.

But isn't this what "2 + 3 = 5" really *says*? That any two things combined with any three more (combined in the right kind of way so that no things are created or destroyed in the process) will make five things? If we *only* understood "2 + 3 = 5" as a quick way of writing a general claim of the latter sort, then again nominalist philosophers of mathematics would not worry.[1] But "2 + 3 = 5" as a mathematical claim is more than a mere abbreviation of a generalization about counting. That this is so can be seen in the fact that it has logical consequences that are not consequences of the generalization to which it relates. It follows logically from "2 + 3 = 5" that there is an object (namely, 2), which, added to 3 makes 5. And this is not a logical consequence of the general claim "For any F, if there are two Fs, and three more Fs, then there are five Fs." For this general claim can be true in finite domains consisting entirely of physical objects, with no numbers in them at all. Since nominalist philosophers question whether there are any numbers (on the grounds that, were there to be such things, they would have to be abstract—nonspatiotemporal, acausal, mind- and language-independent—to serve as appropriate truthmakers for the claims of standard mathematics), they see fit to question claims such as "2 + 3 = 5" precisely because they logically imply the existence of objects such as the number 2, which, they take it, may fail to exist (as in our finite domain example) even though the general claim "any two things added to any three further things make five things" is true.

Some philosophers—inspired by the philosopher/logician Gottlob Frege—try to rule out such finite domains by arguing that the existence of the natural numbers is a consequence of an analytic (or conceptual) truth, this truth being the claim that, effectively, if the members of two collections can be paired off with one another exactly, then they share the same number:

for any F and G, the number of Fs = the number of Gs if and only if $F \approx G$

(where "$F \approx G$" is short for "the Fs and Gs can be put into one–one correspondence"). This claim (which has become known as "Hume's principle," since Frege first presented the principle citing its occurrence in Book 1 of David Hume's *Treatise of Human Nature* [1739]), is argued to be analytic of our concept of number since anyone who grasps the concept of number will grasp the truth of this claim.

Since the existence of numbers falls directly out of Hume's principle, if Hume's principle is part of our concept of number, then it follows from this that anyone who grasps that concept thereby grasps that numbers exist. This derivation of the existence of numbers from our concept of number is reminiscent of St. Anselm's ontological argument, deriving the existence of God from our concept of God as a being "than which no greater can be conceived." (Such a being couldn't exist *merely* in the imagination, Anselm argues, because if we can conceive of God at all, then we can also conceive of Him existing *in reality*. And since existing in reality is greater than existing merely in the imagination, if God existed *only* in the imagination, we could conceive of something even greater—a really existing God—contradicting our definition of God as a being "than which no greater can be conceived.") For nominalist philosophers, the Fregeans' derivation of the existence of numbers from our concept of number is at least as fishy as this supposed derivation of the existence of God from our concept of God. Since nominalist philosophers take themselves to have a concept of number without believing in the existence of numbers, they will reject Hume's principle as a conceptual truth, believing only that Hume's principle characterizes our concept of number in the sense that, *in order for any objects to count as satisfying that concept*, then Hume's principle would have to be true of them, while remaining agnostic on the question of whether there are in fact any numbers.

But *why* remain agnostic about whether there are numbers? And what even hinges on this? Mathematicians talk about mathematical objects and mathematical truths all the time, and indeed are able to *prove* that their mathematical theorems are true. Isn't it absurd in the face of accepted mathematical practice to say, "I know you *think* you've proved Fermat's Last Theorem, Prof. Wiles, but actually since we have no reason to believe there are any numbers, we have no reason to believe FLT"? (Actually, the situation is even worse than that: If there are no numbers, then FLT is trivially true, since it follows a fortiori that there are no numbers $n > 2$ and x, y, $z > 0$, such that $x^n + y^n = z^n$, so Wiles' efforts were truly wasted.) The philosopher David Lewis certainly thought it would be absurd for philosophers to question the truth of mathematical claims. As he puts it,

> Mathematics is an established, going concern. Philosophy is as shaky as can be. To reject mathematics for philosophical reasons would be absurd . . . .
>
> That's not an argument, I know. Rather, I am moved to laughter at the thought of how *presumptuous* it would be to reject mathematics for philosophical reasons. How would *you* like the job of telling the mathematicians that they must change their ways, and abjure countless errors, now that *philosophy* has discovered that

there are no classes? Can you tell them, with a straight face, to follow philosophical argument wherever it may lead? If they challenge your credentials, will you boast of philosophy's other great discoveries: that motion is impossible, that a Being than which no greater can be conceived cannot be conceived not to exist, that it is unthinkable that there is anything outside the mind, that time is unreal, that no theory has ever been made at all probable by the evidence (but on the other hand that an empirically ideal theory cannot possibly be false), that it is a wide-open scientific question whether anyone has ever believed anything, and so on, and on, *ad nauseam*?

Not me!

(Lewis 1991: 58–59)

Just to put this in some perspective, David Lewis is the philosopher best known for believing that, for every true claim about what's possible (such as, "possibly, Trump will win"), there's a world *just like our own* in respect of its reality (i.e., physical, concrete, though spatiotemporally inaccessible to us) at which that claim is actual (i.e., at that world, there is a counterpart to our own Donald Trump, who becomes President of that world's counterpart to our United States).[2] If a philosophical view is so absurd that *even David Lewis can't stomach it*, then maybe it's time to rethink.

Well, if nominalist philosophers are going to find mathematics wanting in the way Lewis suggests (calling on mathematicians to renounce their errors and change their practices), and indeed if as suggested earlier they're going to have to dismiss important results such as Wiles's proof, then they probably do deserve to be laughed out of town. But contemporary nominalists typically wish to leave mathematics just as it is. Indeed, nominalists can even preserve mathematicians' judgments concerning the truth of their theories and the existence of mathematical objects, in at least this sense: there is a notion of truth internal to mathematics according to which to be true *mathematically* just is to be an axiom or a logical consequence of accepted (minimally, logically possible—or *coherent*) mathematical axioms, and to exist *mathematically* just is to be said to exist in an accepted (minimally, logically possible) mathematical theory. Thus, in expressing his puzzlement over Frege's account of axioms in mathematics as truths that are true of all intuitively grasped subject matter, David Hilbert writes in response to a letter from Frege:

> You write: ". . . From the truth of the axioms it follows that they do not contradict one another." I found it very interesting to read this very sentence in your letter. For as long as I have been think-ing and writing on these things, I have been saying the exact re-verse: if the arbitrarily given axioms do not contradict each other in all their consequences, then they are true and the things de-fined by the axioms exist. This is for me the criterion of truth and existence.
>
> Hilbert, letter to Frege, 1899 (reprinted in Frege 1980)

If by truth in mathematics we just mean "axiom or logical consequence of a coherent axiom system," and *if* by existence in mathematics we just mean "existence claim that follows logically from the assumptions of a coherent axiom system," then again the nominalist philosopher will not balk at the *mathematical* truth of the theorems of standard mathematics, or the *mathematical* existence claims that follow from these theorems. Mathematicians are welcome to the truth of their theorems and the existence of mathematical objects, in this sense.[3]

But then what is it that nominalist philosophers *do* balk at? In what sense of truth and existence do they wish to say that we have no reason to believe that the claims of standard mathematics are true or that their objects exist? If we agree that $2 + 3 = 5$ is true in this Hilbertian sense (of being a consequence of coherent axioms), and also true in a practi-cal applied sense (when understood as shorthand for a generalization about what you get when you combine some things and some other things), then what is the nominalist worrying about when she worries whether this sentence is *really* true or whether its objects *really* exist? The issue arises because being true "in the Hilbertian sense" is not al-ways enough. At least, outside of pure mathematics, the mere internal coherence of a framework of beliefs is not enough to count those beliefs as true. Perhaps the notion of an omniscient, omnipotent, omnibenevo-lent being is *coherent*, in the sense that the existence of such a being is at least logically possible, but most would think that there remains a further question as to whether there really is a being satisfying that description. And, in more down-to-Earth matters, Newtonian gravita-tional theory is internally coherent, but we now no longer believe it to be a true account of reality. Granted this general distinction between the mere internal coherence of a theory and its truth, the question arises as to whether we ever have to take our mathematical theories as more than merely coherent—as getting things right about an independently given subject matter. To answer this, we need to understand how we do mathematics—how mathematical theories are developed and applied—and ask whether anything in those practices requires us to say that mathematics is true in anything more than what I have been calling the Hilbertian sense.[4]

It is here that recent debate in the philosophy of mathematics has turned its attention to the role of mathematics in empirical scientific theorizing. Of course, even in unapplied mathematics, *mere* coherence isn't enough.[5] Mathematicians are concerned with developing mathematically interesting theories, axiom systems that are not *merely* coherent but which capture intuitive concepts or have mathematically fruitful consequences. But accounting for the role of these further desiderata does not seem to require that we think of our mathematical theories in the way the Platonist does as answerable to how things really are with a realm of mathematical objects (even if there were such objects, what grounds would we have for thinking that the truths about them should be intuitive, interesting, or fruitful . . . ?). When we turn to the role of mathematics in science, we have at least a prima facie case for taking more than the mere logical possibility of our applied mathematical theories to be confirmed. In particular, close attention has been paid to the alleged explanatory role played by mathematical entities in science. We believe in unobservable theoretical objects such as electrons in part because they feature in the best explanations of observed phenomena: if we explain the appearance of a track in a cloud chamber as having been caused by an electron, but go on to add "but I don't believe in electrons," we seem to undermine our claim to have explained the phenomenon of the track. The same, say many Platonist philosophers of mathematics, goes for mathematical objects such as numbers. If we explain the length of cicada periods (Baker 2005, see also Mark Colyvan's recent paper in *The Mathematical Intelligencer*) as the optimal adaptive choice "because 13 and 17 are prime numbers," but then go on to add, "but I don't believe in numbers," don't we similarly undermine the explanation we have tried to give? On behalf of the nominalist side in this debate, I have argued elsewhere that while mathematics is playing an explanatory role in such cases, it is not mathematical objects that are doing the explanatory work. Rather, such explanations, properly understood, are structural explanations: they explain by showing (a) what would be true in any system of objects satisfying our structure-characterizing mathematical axioms, and (b) that a given physical system satisfies (or approximately satisfies) those axioms. It is because the (axiomatically characterized) natural number structure is instantiated in the succession of summers starting from some first summer at which cicadas appear that the theorem about the optimum period lengths to avoid overlapping with other periods being prime applies. But making use of this explanation does not require any abstract mathematical objects satisfying the Peano axioms, but only that they are true (at least approximately—idealizing somewhat to paper over the fact of the eventual destruction of the Earth) when interpreted as about the succession of summers.

The debate over whether the truth of mathematics, and the existence of mathematical objects (over and above the *Hilbert-truth* and *Hilbert-existence* that comes with mere coherence) is confirmed by the role of mathematics in empirical science, rumbles on. But note that whatever philosophers of science conclude about this issue, it does not impinge on mathematicians continuing to do mathematics as they like, and indeed continuing to make assertions about the (Hilbert)-truth of their theorems and the (Hilbert)-existence of their objects. Nominalists will claim that Hilbert-truth and Hilbert-existence is all that matters when it comes to mathematics, and in this sense it is perfectly fine to agree that $2 + 3 = 5$ (since this is a logical consequence of the Peano axioms). And they will agree that this particular axiom system is of particular interest to us because of the relation of its formally provable claims to logically true generalizations ("If you have two things and three more things, then you have five things"). But to the extent that it is the more-than-mere-coherence *literal* truth of mathematics as a body of claims about a domain of abstract objects that philosophers are concerned about, whereas nominalists may worry whether we have any reason to believe that mathematical claims are true in that sense, perhaps mathematicians can be happy with Russell's definition of mathematics as "the subject in which we never know what we are talking about, nor whether what we are saying is true" (Russell (1910), 58).

Acknowledgments

I am grateful to James Robert Brown, Mark Colyvan, and an anonymous referee of this journal for helpful comments.

Notes

1. Indeed, it is because of the relation of provable-in-PA claims such as "$(1 + 1) + ((1 + 1) + 1) = ((((1 + 1) + 1) + 1) + 1)$" (abbreviated, once suitable definitions are in place, to "$2 + 3 = 5$"), to true logically true generalizations such as "any two things combined with any three other things make five things" that we are interested in the Peano axioms in the first place. A mathematical Platonist—i.e., a defender of the view that mathematics consists of a body of truths about abstract mathematical objects—might say that, far from believing that "$2 + 3 = 5$" on the basis of its following from the Peano axioms, we come to see that the Peano axioms correctly characterize the natural numbers on account of their implying claims, such as the claim that "$2 + 3 = 5$," which we already know to be true of the natural numbers (something like this line of thinking is suggested by Russell's (1907) paper, "The Regressive Method of Discovering the Premises of Mathematics"). The contention of this article is that this line of reasoning is incorrect, since we do not know "$2 + 3 = 5$" to be true of numbers considered as mathematical objects (since we do not know that there are any such objects). Nevertheless, we can mirror this reasoning from an anti-Platonist perspective to provide a justification for PA over other candidate axiom systems: we choose to work on *this* system and are interested in what follows from *its* axioms, in no small part because of the relation of its quantifier-free theorems to logical truths such as the truth that "For any F, if there are two Fs, and three more Fs, then there are five Fs."

2. Typically, in presenting Lewis's account, people appeal to one of his own colorful examples (from Lewis 1986) of a merely possible but nevertheless improbable and in fact nonactual world, for example, a world in which there are talking donkeys. At time of writing (March 2016), I thought I would pick an alternative possible but surely similarly improbable (and thus pretty likely to be nonactual) scenario, one in which Donald Trump becomes President. Lewis's account has as a consequence that whatever the actual outcome of the 2016 election, there would be some possible world in which Trump won it. What I wasn't banking on was that that world would turn out to have been our own.

3. Those familiar with Hilbert's work know that, in his correspondence with Frege, Hilbert would have assumed a syntactic notion of logical consequence, so, strictly speaking, his criterion of truth and existence was *deductive consistency* (so that an axiomatic theory would be true, mathematically speaking, if no contradiction could be derived from those axioms). In light of Gödel's incompleteness theorems, we now know that if we take the second-order Peano axioms (with the full second-order induction axiom, rather than a first-order axiom scheme) and conjoin with this the negation of the Gödel sentence for this theory (defined in relation to a particular derivation system for it), no contradiction will be derivable from this theory, but nevertheless the theory has no model (in the standard second-order semantics). The syntactic notion of deductive consistency thus comes apart (in second-order logic) from the semantic notion of *logically possibly true*. I have used Stewart Shapiro's (1997) terminology of "coherence" as opposed to "consistency" to indicate this semantic notion of logically possible truth. This notion is adequately modeled in mathematics by the model theoretic notion of satisfiability, though I take the lesson of Georg Kreisel (1967) to be that the intuitive notion

of logically possible truth is neither model theoretic nor proof theoretic (though adequately modeled by the model theoretic notion).

4. It is worth noting that Hilbert did not stick with his position that noncontradictoriness is all that is required for truth in mathematics, choosing in his later work to interpret the claims of finitary arithmetic as literal truths about finite strings of strokes (thus straying from his original position that saw axioms as implicit definitions of mathematical concepts, potentially applicable to multiple systems of objects). This later, also Hilbertian, sense of truth (truth when interpreted as claims about syntactic objects), is not the one I wish to advocate in this discussion.

5. It should be noted that, in speaking of "mere" coherence, I do not mean to suggest that establishing the logical possibility of an axiom system is a trivial matter. Substantial work goes into providing relative consistency proofs, and of course the consistency—and so, *a fortiori* coherence—of base theories such as Zermelo–Fraenkel set theory is something about which there is active debate.

References

Alan Baker (2005), "Are There Genuine Mathematical Explanations of Physical Phenomena?" *Mind* 114: 223–238.

James Robert Brown (2017), "Proofs and Guarantees," *The Mathematical Intelligencer* 39(4): 47–50.

Mark Colyvan (2018), "The Ins and Outs of Mathematical Explanation," *The Mathematical Intelligencer* 40(4): 26-9.

Gottlob Frege (1980), *Philosophical and Mathematical Correspondence*, G. Gabriel et al., eds., and B. McGuinness, trans. (Oxford, U.K.: Blackwell).

David Hume (1739), *Treatise of Human Nature*, D. F. Norton and M. J. Norton, eds. (Oxford, U.K.: Oxford University Press).

Georg Kreisel (1967), "Informal Rigour and Completeness Proofs," in Imre Lakatos, ed., *Problems in the Philosophy of Mathematics* (Amsterdam, Netherlands: North-Holland): 138–171.

David Lewis (1986), *On the Plurality of Worlds* (Oxford, U.K.: Blackwell).

David Lewis (1991), *Parts of Classes* (Oxford, U.K.: Blackwell).

Bertrand Russell (1907), "The Regressive Method of Discovering the Premises of Mathematics," in his *Essays in Analysis* (London: George Allen & Unwin Ltd., 1973), 272–283.

Bertrand Russell (1910), "Mathematics and the Metaphysicians," in his *Mysticism and Logic and Other Essays* (London: George Allen & Unwin Ltd., 1917).

Stewart Shapiro (1997), *Philosophy of Mathematics: Structure and Ontology* (Oxford, U.K.: Oxford University Press).

Gregory's Sixth Operation

Tiziana Bascelli, Piotr Błaszczyk,
Vladimir Kanovei, Karin U. Katz,
Mikhail G. Katz, Semen S. Kutateladze,
Tahl Nowik, David M. Schaps, and David Sherry

1. Introduction

Marx Wartofsky pointed out in his programmatic contribution *The Relation between Philosophy of Science and History of Science* that there are many distinct possible relations between philosophy of science and history of science, some "more agreeable" and fruitful than others (Wartofsky 1976, p. 719ff). Accordingly, a fruitful relation between history and philosophy of science requires a rich and complex *ontology* of that science. In the case of mathematics, this means that a fruitful relation between history and philosophy must go beyond offering an ontology of the domain over which a certain piece of mathematics ranges (say, numbers, functions, sets, infinitesimals, or structures). Namely, it must develop the ontology of mathematics *as a scientific theory* itself (Wartofsky 1976, p. 723). A crucial distinction here is that between the (historically relative) *ontology* of the mathematical objects in a certain historical setting and its *procedures*, particularly emphasizing the different roles these components play in the history of mathematics. More precisely, *procedures* serve as a representative of what Wartofsky called the *praxis* characteristic of the mathematics of a certain time period, and *ontology* in the narrow sense takes care of the mathematical entities recognized at that time. On the procedure/entity distinction, A. Robinson had this to say:

> ... from a formalist point of view we may look at our theory syntactically and may consider that what we have done is to introduce *new deductive procedures* rather than new mathematical entities.
>
> (Robinson 1966, p. 282) (emphasis in the original)

As a case study, we analyze the text *Vera Circuli* (Gregory 1667) by James Gregory.

2. Ultimate Terms and Termination of Series

Gregory studied under Italian indivisibilists[1] and specifically Stefano degli Angeli during his years 1664–1668 in Padua. Some of Gregory's first books were published in Italy. His mathematical accomplishments include the series expansions not only for the sine but also for the tangent and secant functions (González-Velasco 2011).

The *Vera Circuli* contains a characterization of the "termination" of a convergent *series* (i.e., *sequence* in modern terminology). This sequence was given by Gregory in the context of a discussion of a double sequence (lower and upper bounds) of successive polygonal approximations to the area of a circle:

> & igitur imaginando hanc seriem in infinitum continuari, possimus imaginari vltimos terminos couergentes [sic] esse equales, quos terminos equales appellamus seriei terminationem.
>
> (Gregory 1667, pp. 18–19)

In the passage above, Gregory's *seriem* refers to a *sequence*, and the expression *terminus* has its usual meaning of a *term* of a sequence. The passage can be rendered in English as follows:

> And so by imagining this series [i.e., sequence] to be continued to infinity, we can imagine the ultimate convergent terms *to be equal*; and we call those equal ultimate terms the termination of the series. [emphasis added]

Lützen (2014, p. 225) denotes the lower and upper bounds respectively by I_n (for *inscribed*) and C_n (for *circumscribed*). Gregory proves the recursive formulas $I_{n+1}^2 = C_n I_n$ and $C_{n+1} = \frac{2C_n I_{n+1}}{C_n + I_{n+1}}$. Gregory states that the "ultimate convergent terms" of the sequences I_n and C_n are *equal*.

After having defined the two series of inscribed and circumscribed polygons, Gregory notes:

> atque in infinitum illam [=hanc polygonorum seriem] continuando, manifestum est tandem exhiberi quantitatem sectori circulari, elliptico vel hyperbolico ABEIOP æquale[m]; differentia enim

polygonorum complicatorum in seriei continuatione semper dimi-
nuitur, ità vt omni exhibita quantitate fieri possit minor, & vt in
sequenti theorematis Scholio demonstrabimus: si igitur prædicta
polygonorum series terminari posset, hoc est, si inueniretur vlti-
mum illud polygonum inscriptum (si ità loquì liceat) æquale vltimo
illi polygono circumscripto, daretur infallibiliter circuli & hyper-
bolæ quadratura: sed quoniam difficile est, & in geometria omninò
fortasse inauditu[m] tales series terminare; præmitte[n]dæ sunt
quæ dam propositiones è quibus inueniri possit huiusmodi aliquot
serierum terminationes, & tandem (si fieri possit) generalis methodus inueniendi omnium serierum co[n]uergentium terminationes.

This can be translated as follows:

and that [series of polygons] being continued to infinity, it is clear
that a quantity equal to a circular, elliptic, or hyperbolic sector
ABEIOP will be produced. The difference between [two *n*th terms]
in the continuation of the series of complicated polygons always di-
minishes so that it can become less than any given quantity indeed,
as we will prove in the Scholium to the theorem. Thus, if the above-
mentioned series of polygons can be terminated, that is, if that ultimate inscribed polygon is found to be equal (so to speak) to that
ultimate circumscribed polygon, it would undoubtedly provide the
quadrature of a circle as well as a hyperbola. But since it is difficult,
and in geometry perhaps unheard-of, for such a series to come to an
end [lit.: be terminated], we have to start by showing some propositions by means of which it is possible to find the terminations of a
certain number of series of this type, and finally (if it can be done)
a general method of finding terminations of all convergent series.

The passage clearly shows that Gregory is using the term "ultimate (or
last) circumscribed polygon" in a figurative sense, as indicated by

- his parenthetical "so to speak," which indicates that he is not
  using the term literally; and
- his insistence that "in geometry it is unheard-of" for a sequence
  to come to be terminated.

He makes it clear that he is using the word *termination* in a new sense,
which is precisely his sixth operation, as discussed below.

One possible interpretation of *ultimate terms* would be the following. This could refer to those terms that are all closer than epsilon to one another. If ordinary terms are *further* than epsilon, that would make them different. The difficulty for this interpretation is that, even if ordinary terms are *closer* than epsilon, they are still *different*, contrary to what Gregory wrote about their being *equal*. M. Dehn and E. Hellinger attribute to Gregory

> a very general, new analytic process which he coordinates as the "sixth" operation along with the five traditional operations (addition, subtraction, multiplication, division, and extraction of roots). In the introduction, he proudly states "ut hae c nostra inventio addat arithmeticae aliam operationem et geometriae aliam rationis speciem, ante incognitam orbi geometrico." This operation is, as a matter of fact, our modern limiting process.
>
> (Dehn and Hellinger 1943, pp. 157–158)

We will have more to say about what this sixth operation could be *as a matter of fact* (see Section 4 on shadow-taking). A. Malet expressed an appreciation of Gregory's contribution to analysis in the following terms:

> Studying Gregorie's work on "Taylor" expansions and his analytical method of tangents, which has passed unnoticed so far, [we argue] that Gregorie's work is a counter-example to the standard thesis that geometry and algebra were opposed forces in 17th-century mathematics.
>
> (Malet 1989, p. 1)

What is, then, Gregory's *sixth operation* mentioned by Dehn and Hellinger, and how is it related to convergence?

3. Law of Continuity

The use of infinity was not unusual for this period. As we mentioned in the introduction, Gregory fit naturally in the proud Italian tradition of the method of indivisibles, and was a student of Stefano degli Angeli at Padua between 1664 and 1668. Degli Angeli published sharp responses to critiques of indivisibles penned by Jesuits Mario Bettini and André Tacquet. Bettini's criticisms were extensions of earlier criticisms by

Jesuit Paul Guldin. Degli Angeli defended the method of indivisibles against their criticisms.

Both indivisibles and degli Angeli himself appear to have been controversial at the time in the eyes of the Jesuit order, which banned indivisibles from being taught in their colleges on several occasions. Thus, in 1632 (the year Galileo was summoned to stand trial over heliocentrism) the Society's Revisors General (this "Society" refers to the Jesuit order) led by Jacob Bidermann banned teaching indivisibles in their colleges (Festa 1990, 1992, p. 198). Indivisibles were placed on the Society's list of *permanently* banned doctrines in 1651 (Hellyer 1996).

It seems that Gregory's 1668 departure from Padua was well timed, for his teacher degli Angeli's Jesuate order[2] was suppressed by papal brief in the same year, cutting short degli Angeli's output on indivisibles. Gregory's own books were suppressed at Venice, according to a letter from John Collins to Gregory dated November 25, 1669, in which he writes

> One Mr. Norris a Master's Mate recently come from Venice, saith it was there reported that your bookes were suppressed, not a booke of them to be had anywhere, but from Dr. Caddenhead to whom application being made for one of them, he presently sent him one (though a stranger) refusing any thing for it.
>
> (Turnbull 1939, p. 74)

In a 1670 letter to Collins, Gregory writes

> I shall be very willing ye writ to Dr Caddenhead in Padua, for some of my books. In the mean time, I desire you to present my service to him, and to inquire of him if my books be suppressed, and the reason thereof.
>
> (Gregory to Collins, St. Andrews,
> March 7, 1670, in Turnbull p. 88)

In a letter to Gregory, written in London, September 29, 1670, Collins reported as follows: "Father Bertet[3] sayth your Bookes are in great esteeme, but not to be procured in Italy" (Turnbull p. 107).

The publishers' apparent reluctance to get involved with Gregory's books may also explain degli Angeli's silence on indivisibles following the suppression of his order, but it is hard to say anything definite in the matter until the archives at the Vatican dealing with the suppression of the Jesuates are opened to independent researchers. Certainly one can understand Gregory's own caution in matters infinitesimal (of course, the latter term wasn't coined until later).

John Wallis introduced the symbol ∞ for an infinite number in his *Arithmetica Infinitorum* (Wallis 1656) and exploited an infinitesimal number of the form $\frac{1}{\infty}$ in area calculations (Scott 1981, p. 18), more than a decade before the publication of Gregory's *Vera Circuli*. At about the same time, Isaac Barrow "dared to explore the logical underpinnings of infinitesimals," as Malet put it:

> Barrow, who dared to explore the logical underpinnings of infinitesimals, was certainly modern and innovative when he publicly defended the new mathematical methods against Tacquet and other mathematical "classicists" reluctant to abandon the Aristotelian continuum. And after all, to use historical hindsight, it was the non-Archimedean structure of the continuum linked to the notion of infinitesimal and advocated by Barrow that was to prove immensely fruitful as the basis for the Leibnizian differential calculus.
>
> (Malet 1989, p. 244)

We know that G. W. Leibniz was an avid reader of Gregory; see e.g., Leibniz (1672). To elaborate on the link to Leibniz mentioned by Malet, note that Leibniz might have interpreted Gregory's definition of convergence as follows. Leibniz's *law of continuity* (Leibniz 1702, pp. 93–94) asserts that whatever succeeds in the finite succeeds also in the infinite, and vice versa; see Katz and Sherry (2013) for details. Thus, if one can take terms of a sequence corresponding to a finite value of the index n, one should also be able to take terms corresponding to infinite values of the index n. What Gregory refers to as the "ultimate" terms would then be the terms I_n and C_n corresponding to an infinite index n.

Leibniz interpreted equality as a relation in a larger sense of equality *up to* (negligible terms). This was codified as his *transcendental law of homogeneity* (Leibniz 1710); see Bos (1974, p. 33) for a thorough discussion. Thus, Leibniz wrote

> Caeterum aequalia esse puto, non tantum quorum differentia est omnino nulla, sed et quorum differentia est incomparabiliter parva; et licet ea Nihil omnino dici non debeat, non tamen est quantitas comparabilis cum ipsis, quorum est differentia.
>
> (Leibniz 1695, p. 322)

This can be translated as follows:

> Furthermore, I think that not only those things are equal whose difference is absolutely zero, but also whose difference is incomparably small. And although this [difference] need not absolutely be called Nothing, neither is it a quantity comparable to those whose difference it is.

In the seventeenth century, such a generalized notion of equality was by no means unique to Leibniz. Indeed, Leibniz himself cites an antecedent in Pierre de Fermat's technique (known as the method of *adequality*), in the following terms:

> Quod autem in aequationibus Fermatianis abjiciuntur termini, quos ingrediuntur talia quadrata vel rectangula, non vero illi quos ingrediuntur simplices lineae infinitesimae, ejus ratio non est quod hae sint aliquid, illae vero sint nihil, sed quod termini ordinarii per se destruuntur.[4]
>
> (Leibniz 1695, p. 323)

On this page, Leibniz describes Fermat's method in a way similar to Leibniz's own. On occasion Leibniz used the notation "⊓" for the relation of equality. Note that Leibniz also used our symbol "=" and other signs for equality and did not distinguish between "=" and "⊓" in this regard. To emphasize the special meaning *equality* had for Leibniz, it may be helpful to use the symbol "⊓" so as to distinguish Leibniz's equality from the modern notion of equality "on the nose." Then Gregory's comment about the equality of the ultimate terms translates into

$$I_n \sqcap C_n(1)$$

when n is infinite.

From the viewpoint of the modern Weierstrassian framework, it is difficult to relate to Gregory's insight. Thus, G. Ferraro translates Gregory's "vltimos terminos conuergentes" as "last convergent terms" (Ferraro 2008, p. 21), and goes on a few pages later to mention

> Gregory's reference to *the last term*, p. 21. . . . In Leibniz they appear in a clearer way.
>
> (Ferraro 2008, p. 27, note 41) (emphasis added)

Ferraro may have provided an accurate translation of Gregory's comment, but Ferraro's assumption that there is something unclear about

Gregory's comment because of an alleged "last term" is unjustified. Note that Ferraro's use of the singular "last term" (note 41) is not consistent with Gregory's use of the plural *terminos* (terms) in his book. One may find it odd for a mathematician of Gregory's caliber to hold that there is literally a *last* term in a sequence. Dehn and Hellinger mention only the plural "last convergent terms" (Dehn and Hellinger 1943, p. 158).

4. The Unguru Controversy

There is a debate in the community of historians whether it is appropriate to use modern theories and/or modern notation in interpreting mathematical texts of the past, with S. Unguru a staunch opponent, whether with regard to interpreting Euclid, Apollonius, or Fermat (Unguru 1976). See Corry (2013) for a summary of the debate. Note that Ferraro does not follow Unguru in this respect. Indeed, Ferraro exploits the modern notation

$$\sum_{i=1}^{\infty} a_i \tag{2}$$

for the sum of the series, already on page 5 of his book, while discussing late sixteenth-century texts of Viète. We note the following two aspects of the notation, as seen in Equation (2):

(1) It presupposes the modern epsilontic notion of limit, where $S = \sum_{i=1}^{\infty} a_i$ means $\forall \varepsilon > 0 \, \exists N \in \mathbb{N} \, (n > N \Rightarrow \left| S - \sum_{i=1}^{n} a_i \right| < \varepsilon)$ in the context of a Weierstrassian framework involving a strictly Archimedean punctiform continuum;

(2) The symbol "∞" occurring in Ferraro's formula has no meaning other than a reminder that a limit was taken in the construction. In particular, this usage of the symbol ∞ is distinct from its original seventeenth-century usage by Wallis, who used it to denote a specific infinite number, and proceeded to work with infinitesimal numbers like $\frac{1}{\infty}$ (see Section 3).

We will avoid choosing sides in the debate over Unguru's proposal.[5] However, once one resolves to exploit modern frameworks involving punctiform continua and/or number systems, as Ferraro does, to interpret seventeenth-century texts, one still needs to address the following important question:

Which modern framework is more appropriate for interpreting the said historical texts?

Here appropriateness could be gauged in terms of providing the best proxies for the *procedural* moves found in the great seventeenth-century masters.

Hacking (2014) points out that there is a greater amount of contingency in the historical evolution of mathematics than is generally thought. Hacking proposes a *Latin model* of development (of a natural language like Latin, with the attendant contingencies of development caused by social factors) to the usual *butterfly model* of development (of a biological organism like a butterfly, which is genetically predetermined in spite of apparently discontinuous changes in its development). This model tends to undercut the apparent inevitability of the Weierstrassian model.

We leave aside the *ontological* or foundational questions of how to justify the entities like points or numbers (in terms of modern mathematical foundations) and focus instead on the *procedures* of the historical masters, as discussed in Section 1. More specifically, is a modern Weierstrassian framework based on an Archimedean continuum more appropriate for interpreting their procedures, or is a modern infinitesimal system more appropriate for this purpose?

Note that in a modern infinitesimal framework such as Robinson's, sequences possess terms with infinite indices. Gregory's relation can be formalized in terms of the standard part principle in Robinson's framework (Robinson 1966). This principle asserts that every finite hyperreal number is infinitely close to a unique real number.

In more detail, in a hyperreal extension $\mathbb{R} \hookrightarrow {}^*\mathbb{R}$ one considers the set ${}^h\mathbb{R} \subseteq {}^*\mathbb{R}$ of *finite* hyperreals. The set ${}^h\mathbb{R}$ is the domain of the standard part function (also called the *shadow*) $\mathbf{st} : {}^h\mathbb{R} \to \mathbb{R}$, rounding off each finite hyperreal number to its nearest real number.

In the world of James Gregory, if each available term with an infinite index n is indistinguishable (in the sense of being infinitely close) from some standard number, then we "terminate the series" (to exploit Gregory's terminology) with this number, meaning that this number is the limit of the sequence. Gregory's definition corresponds to a relation of infinite proximity in a hyperreal framework. Namely we have

$$I_n \approx C_n \tag{3}$$

where ≈ is the relation of being infinitely close (i.e., the difference is infinitesimal), and the common standard part of these values is the limit of the sequence. Equivalently, $\mathbf{st}(I_n) = \mathbf{st}(C_n)$. Mathematically speaking, this is equivalent to a Weierstrassian epsilontic paraphrase along the lines of item (1) above.

Recently Robinson's framework has become more visible, thanks to high-profile advocates like Terry Tao (Tao 2014, Tao and Vu 2016). The field has also had its share of high-profile detractors, such as Errett Bishop and Alain Connes. Their critiques were critically analyzed in Katz and Katz (2011), Katz and Leichtnam (2013), Kanovei et al. (2013), and Sanders (2017). Further criticisms by J. Earman, K. Easwaran, H. M. Edwards, G. Ferraro, J. Gray, P. Halmos, H. Ishiguro, G. Schubring, and Y. Sergeyev were dealt with, respectively, in the following recent texts: Katz and Sherry (2013), Bascelli et al. (2014, 2016), Kanovei et al. (2015), Bair et al. (2017), Błaszczyk et al. (2016, 2017a, b), and Gutman et al. (2017). In Borovik and Katz (2012), we analyze the Cauchy scholarship of Judith Grabiner. For a fresh look at Simon Stevin, see Katz and Katz (2012).

5. Conclusion

We note a close fit between Gregory's procedure (Equation [1]) and the procedure in Equation (3) available in a modern infinitesimal framework. The claim that "[Gregory's] definition is rather different from the modern one" (Ferraro 2008, p. 20) is only true with regard to a *Weierstrassian* modern definition. Exploiting the richer syntax available in a modern infinitesimal framework where Gregory's procedure acquires a fitting proxy, it is possible to avoid the pitfalls of attributing to a mathematician of Gregory's caliber odd beliefs in an alleged "last" term in a sequence.

An infinitesimal framework also enables an interpretation of the notion of "ultimate terms" as proxified by terms with infinite index, and "termination of the series" as referring to the assignable number infinitely close to a term with an infinite index, by Leibniz's transcendental law of homogeneity (or the standard part principle of Robinson's framework).

Whereas some scholars seek to interpret Gregory's procedures in a default modern post-Weierstrassian framework, arguably a modern infinitesimal framework provides better proxies for Gregory's procedural moves than a modern Weierstrassian one.

Acknowledgments

M. Katz was partially supported by the Israel Science Foundation Grant No. 1517/12.

Notes

1. Today scholars distinguish carefully between indivisibles (i.e., codimension one objects) and infinitesimals (i.e., of the same dimension as the entity they make up); see e.g., Koyré (1954). However, in the seventeenth century the situation was less clearcut. The term *infinitesimal* itself was not coined until the 1670s; see Katz and Sherry (2013).

2. This was an older order than the Jesuits. Cavalieri had also belonged to the Jesuate order.

3. Jean Bertet (1622–1692), Jesuit, quit the order in 1681. In 1689, Bertet conspired with Leibniz and Antonio Baldigiani in Rome to have the ban on Copernicanism lifted (Wallis 2012).

4. Translation: "But the fact that in Fermat's equations those terms into which such things enter as squares or rectangles [i.e., multiplied by themselves or by each other] are eliminated but not those into which simple infinitesimal lines [i.e., segments] enter—the reason for that is not because the latter are something whereas the former are really nothing [as Nieuwentijt maintained], but because ordinary terms cancel each other out."

5. The sources of such a proposal go back (at least) to A. Koyré who wrote: "Le problème du langage à adopter pour l'exposition des oeuvres du passé est extrêmement grave et ne comporte pas de solution parfaite. En effet, si nous gardons la langue (la terminologie) de l'auteur étudié, nous risquons de le laisser incompréhensible, et si nous lui substituons la nôtre, de le trahir." (Koyré 1954, p. 335, note 3).

References

Bair, J., Błaszczyk, P., Ely, R., Henry, V., Kanovei, V., Katz, K., et al. (2017). Interpreting the infinitesimal mathematics of Leibniz and Euler. *Journal for General Philosophy of Science, 48*(1). doi:10.1007/s10838-016-9334-z, http://arxiv.org/abs/1605.00455.

Bascelli, T., Bottazzi, E., Herzberg, F., Kanovei, V., Katz, K., Katz, M., et al. (2014). Fermat, Leibniz, Euler, and the gang: The true history of the concepts of limit and shadow. *Notices of the American Mathematical Society, 61*(8), 848–864.

Bascelli, T., Błaszczyk, P., Kanovei, V., Katz, K., Katz, M., Schaps, D., et al. (2016). Leibniz vs Ishiguro: Closing a quarter-century of syncategoremania. *HOPOS: Journal of the International Society for the History of Philosophy of Science, 6*(1), 117–147. doi:10.1086/685645, http://arxiv.org/abs/1603.07209.

Błaszczyk, P., Borovik, A., Kanovei, V., Katz, M., Kudryk, T., Kutateladze, S., et al. (2016). A non-standard analysis of a cultural icon: The case of Paul Halmos. *Logica Universalis, 10*(4), 393–405. doi:10.1007/s11787-016-0153-0, http://arxiv.org/abs/1607.00149.

Błaszczyk, P., Kanovei, V., Katz, K., Katz, M., Kutateladze, S., and Sherry. D. (2017a). Toward a history of mathematics focused on procedures. *Foundations of Science.* doi:10.1007/s10699-016-9498-3, http://arxiv.org/abs/1609.04531.

Błaszczyk, P., Kanovei, V., Katz, M., and Sherry, D. (2017b). Controversies in the foundations of analysis: Comments on Schubring's *Conflicts. Foundations of Science.* doi:10.1007/s10699-015-9473-4, http://arxiv.org/abs/1601.00059.

Borovik, A., and Katz, M. (2012). Who gave you the Cauchy–Weierstrass tale? The dual history of rigorous calculus. *Foundations of Science, 17*(3), 245–276. doi:10.1007/s10699-011-9235-x.

Bos, H. (1974). Differentials, higher-order differentials and the derivative in the Leibnizian calculus. *Archive for History of Exact Sciences, 14*, 1–90.

Corry, L. (2013). Geometry and arithmetic in the medieval traditions of Euclid's *Elements*: A view from Book II. *Archive for History of Exact Sciences, 67*(6), 637–705.

Dehn, M., and Hellinger, E. (1943). Certain mathematical achievements of James Gregory. *The American Mathematical Monthly, 50*, 149–163.

Ferraro, G. (2008). *The rise and development of the theory of series up to the early 1820s: Sources and studies in the history of mathematics and physical sciences.* New York: Springer.

Festa, E. (1990). La querelle de l'atomisme: Galilée, Cavalieri et les Jésuites. *La Recherche* (Sept. 1990), 1038–1047.

Festa, E. (1992). Quelques aspects de la controverse sur les indivisibles. Geometry and atomism in the Galilean school, 193–207, Bibl. Nuncius Studi Testi, X, Olschki, Florence.

González-Velasco, E. (2011). *Journey through mathematics: Creative episodes in its history.* New York: Springer.

Gregory, J. (1667). *Vera Circuli et Hyperbolae Quadratura.* Padua edition, 1667. Patavia edition, 1668.

Gutman, A., Katz, M., Kudryk, T., and Kutateladze, S. (2017). *The Mathematical Intelligencer* flunks the olympics. *Foundations of Science, 22*(3), 539–555.

Hacking, I. (2014). *Why is there philosophy of mathematics at all?* Cambridge, U.K.: Cambridge University Press.

Hellyer, M. (1996). "Because the authority of my superiors commands": Censorship, physics and the German Jesuits. *Early Science and Medicine, 3*, 319–354.

Kanovei, V., Katz, K., Katz, M., and Sherry, D. (2015). Euler's lute and Edwards' oud. *The Mathematical Intelligencer, 37*(4), 48–51. doi:10.1007/s00283-015-9565-6, http://arxiv.org/abs/1506.02586.

Kanovei, V., Katz, M., and Mormann, T. (2013). Tools, objects, and chimeras: Connes on the role of hyperreals in mathematics. *Foundations of Science, 18*(2), 259–296. doi:10.1007/s10699-012-9316-5, http://arxiv.org/abs/1211.0244.

Katz, K., and Katz, M. (2011). Meaning in classical mathematics: Is it at odds with intuitionism? *Intellectica, 56*(2), 223–302. http://arxiv.org/abs/1110.5456.

Katz, K., and Katz, M. (2012). Stevin numbers and reality. *Foundations of Science, 17*(2), 109–123. doi:10.1007/s10699-011-9228-9, http://arxiv.org/abs/1107.3688.

Katz, M., and Leichtnam, E. (2013). Commuting and noncommuting infinitesimals. *American Mathematical Monthly, 120*(7), 631–641. doi:10.4169/amer.math.monthly.120.07.631, http://arxiv.org/abs/1304.0583.

Katz, M., and Sherry, D. (2013). Leibniz's infinitesimals: Their fictionality, their modern implementations, and their foes from Berkeley to Russell and beyond. *Erkenntnis, 78*(3), 571–625.

Koyré, A. (1954). Bonaventura Cavalieri et la géométrie des continus. In *Etudes d'histoire de la pensée scientifique*, Gallimard, 1973. Originally published in *Hommage à Lucien Febvre.* Paris: Colin.

Leibniz, G. (1695). Responsio ad nonnullas difficultates a Dn. Bernardo Niewentiit circa methodum differentialem seu infinitesimalem motas. *Act. Erudit. Lips.* (1695). In Gerhardt, C. (Ed.), *Leibnizens mathematische Schriften* (Vol. V, pp. 320–328). Berlin and Halle: Eidmann. A French translation is in [Leibniz 1989, pp. 316–334].

Leibniz, G. (1702). To Varignon, 2 feb. 1702. In Gerhardt, C. (Ed.), *Leibnizens mathematische Schriften* (Vol. IV, pp. 91–95). Berlin and Halle: Eidmann.

Leibniz, G. (1710). Symbolismus memorabilis calculi algebraici et infinitesimalis in comparatione potentiarum et differentiarum, et de lege homogeneorum transcendentali. In Gerhardt, C. (Ed.), *Leibnizens mathematische Schriften* (Vol. V, pp. 377–382). Berlin and Halle: Eidmann.

Leibniz, G. (1989). La naissance du calcul différentiel. 26 articles des *Acta Eruditorum*. Translated from the Latin and with an introduction and notes by Marc Parmentier. With a preface by Michel Serres. Mathesis. Librairie Philosophique J. Vrin, Paris.

Leibniz, G. W. (1672). Sämtliche Schriften und Briefe. Reihe 7. Mathematische Schriften. Band 6. pp. 1673–1676. Arithmetische Kreisquadratur. [Collected works and letters. Series VII. Mathematical writings. Vol. 6, pp. 1673–1676. Arithmetic squaring of the circle] Edited by Uwe Mayer and Siegmund Probst. With an introduction and editorial remarks in German. Akademie Verlag, Berlin, 2012. Vol. VII, 3, no. 6, 65.

Lützen, J. (2014). 17th century arguments for the impossibility of the indefinite and the definite circle quadrature. *Revue d'histoire des mathématiques, 20*(2), 211–251.

Malet, A. (1989). *Studies on James Gregorie (1638–1675)*. Ph.D. thesis, Princeton University.

Robinson, A. (1966). *Non-standard analysis*. Amsterdam, Netherlands: North-Holland.

Sanders, S. (2017). Reverse formalism 16. *Synthese* http://dx.doi.org/10.1007/S11229-017-1322-2 and https://arxiv.org/abs/1701.05066.

Scott, J. (1981). *The mathematical work of John Wallis, D.D., F.R.S. (1616–1703)*. Second ed. With a foreword by E. N. da C. Andrade. New York: Chelsea Publishing.

Tao, T. (2014). *Hilbert's fifth problem and related topics*. Graduate Studies in Mathematics 153. American Mathematical Society, Providence, RI.

Tao, T., and Vu, V. (2016). Sum-avoiding sets in groups. *Discrete Analysis*. doi:10.19086/da.887, http://arxiv. org/abs/1603.03068.

Turnbull, H. (1939). *James Gregory tercentenary memorial volume. Royal Society of Edinburgh*. London: G. Bell and Sons.

Unguru, S. (1976). Fermat revivified, explained, and regained. *Francia, 4*, 774–789.

Wallis, J. (1656). *Arithmetica infinitorum sive Nova Methodus Inquirendi in Curvilineorum Quadraturam, aliaque difficiliora Matheseos Problemata*. Oxonii. Typis Leon Lichfield Academiae Typographi Impensis Tho. Robinson.

Wallis, J. (2012). *The correspondence of John Wallis. Vol. III (October 1668–1671)*. Edited by Philip Beeley and Christoph J. Scriba. Oxford: Oxford University Press.

Wartofsky, M. (1976). The relation between philosophy of science and history of science. In R. S. Cohen, P. K. Feyerabend, and M. W. Wartofsky (Eds.), *Essays in memory of Imre Lakatos* (pp. 717–737), Boston studies in the philosophy of science XXXIX Dordrecht, Netherlands: D. Reidel Publishing.

Kolmogorov Complexity and Our Search for Meaning: What Math Can Teach Us about Finding Order in Our Chaotic Lives

Noson S. Yanofsky

Was it a chance encounter when you met that special someone, or was there some deeper reason for it? What about that strange dream last night—was that just the random ramblings of the synapses of your brain, or did it reveal something deep about your unconscious? Perhaps the dream was trying to tell you something about your future. Perhaps not. Did the fact that a close relative developed a virulent form of cancer have profound meaning, or was it simply a consequence of a random mutation of his DNA?

We live our lives thinking about the patterns of events that happen around us. We ask ourselves whether they are simply random or if there is some reason for them that is uniquely true and deep. As a mathematician, I often turn to numbers and theorems to gain insight into questions like these. As it happens, I learned something about the search for meaning within patterns in life from one of the deepest theorems in mathematical logic. That theorem, simply put, shows that there is no way to know, even in principle, if an explanation for a pattern is the deepest or most interesting explanation there is. Just as in life, the search for meaning in mathematics knows no bounds.

First, some preliminaries. Consider the following three strings of characters:

1. 001001001001001001001001001001001001001001001001001001001100100
2. 2, 3, 5, 7, 11, 13, 17, 19, 23, 29, 31, 37, 43, 47, 53, 59, 61, 67, 71, 73, 79, 83, 89, 97
3. 38386274868783254735796801834682918987459817087106701409581980418

How can we describe these strings? We can easily describe them by just writing them down as we just did. However, it is pretty obvious that there are shorter descriptions of the first two strings. The first is simply the pattern "100" over and over. The second pattern is simply a listing of the first few prime numbers. What about the third string? We can describe it by just printing the string. But is there a better, shorter description?

In the early 1960s, an American teenager named Gregory Chaitin, the world-famous Russian mathematician Andrey Kolmogorov, and the computer science pioneer Ray Solomonoff independently formulated a way of measuring the complexity of strings of characters. Their ideas have come to be called Kolmogorov complexity theory or algorithmic information theory. They posited that a string is as complex as the length of the shortest computer program that can produce the string. That is, take a string and look for a short computer program that produces that string. The program is a type of description of the string. If the shortest such program is very short, then the string has a simple pattern and is not very complex. We say that string "has little algorithmic content." In contrast, if a long program is required to produce the string, then the string is complicated and "has more algorithmic content." For any string, one must look for the shortest program that produces that string. The length of that program is called the Kolmogorov complexity of the string.

Let us look at the above three strings. The first two strings can be described by relatively short computer programs:

1. Print "100" 23 times.
2. Print the first 24 prime numbers.

The Kolmogorov complexity of the first string is less than the Kolmogorov complexity of the second string because the first program is shorter than the second program. What about the third string? There is no obvious pattern for this string. Nevertheless, there exists a silly program that prints this sequence:

3. Print "3838627486878325473579680183468291898745981708710670140958198 0418".

This program will do the job, but it is not satisfying. Perhaps a shorter program shows that the string has a pattern. When the shortest program to produce a string is simply "Print the string," we say that the string is complicated and there is no known pattern. A string that lacks any pattern is called *random*. Although we do not see any pattern, there could still be one. In mathematics, as in life, we are confronted with many seemingly random patterns.

We might try to use the amazing powers of modern computers to find a pattern and a shortest program. Wouldn't it be lovely if there were a computer that would simply calculate the Kolmogorov complexity of any string? This computer would accept a string as input and would output the length of the shortest program that can produce that string. Surely, with all the newfangled computer tools, such as artificial intelligence, deep learning, big data, and quantum computing, it would be easy to create such a computer.

Alas, no such computer can exist! As powerful as modern computers are, this task cannot be accomplished. This is the content of one of the deepest theorems in mathematical logic. Basically, the theorem says that the Kolmogorov complexity of a string cannot be computed. There is no mechanical device to determine the size of the smallest program that produces a given string. It is not that our current level of computer technology is insufficient for the task at hand, or that we are not clever enough to write the algorithm. Rather, it was proven that the very notion of description and computation shows that no such computer can ever possibly perform the task for every string. While a computer might find some pattern in a string, it cannot find the *best* pattern. We might find some short program that outputs a certain pattern, but there could exist an even shorter program. We will never know.

The proof that the Kolmogorov complexity of a sequence is not computable is a bit technical. But it is a proof by contradiction, and we can get a sense of how that works by looking at two cute little paradoxes.

The interesting number paradox revolves around the claim that all natural numbers are interesting. 1 is the first number, so that is interesting. 2 is the first even number. 3 is the first odd prime number. 4 is interesting because $4 = 2 \times 2$ and $4 = 2 + 2$. We can continue in this fashion and find interesting properties for many numbers. At some point, we might come to some number that does not seem to have an interesting property. We can call that number the first uninteresting number. But that, in itself, is an interesting property. In conclusion, the uninteresting number is, in fact, interesting!

The ideas inside the Kolmogorov proof are also similar to those in the Berry paradox, which is about describing large numbers. Notice that the more words you use, the larger the number you can describe. For example, in three words you can describe "a trillion trillion," and in five words you can describe "a trillion trillion trillion trillion," which is much larger. Now consider the number described by the following phrase:

The smallest number that cannot be described in less than 15 words.

This number needs 15, or 16, or even more words to describe it. It cannot be described by 12 words, or 13 words, or 14 words. However there is a major problem: The above phrase described the number in only 12 words. Our description of the number violated the description of the number. This is a contradiction.

In both the interesting number paradox and the Berry paradox, we arrive at contradictions by assuming that there is an exact way of describing something. Similarly, the proof that Kolmogorov complexity is not computable springs from the fact that if it was, we would find a contradiction.

The fact that Kolmogorov complexity is not computable is a result in pure mathematics, and we should never confuse that pristine realm with the far more complicated, and messy, real world. However, there are certain common themes about Kolmogorov complexity theory that we might take with us when thinking about the real world.

Many times we are presented with something that looks totally chaotic. This randomness is unnerving, and so we search for a pattern that eliminates some of the chaos. If we do find a pattern, it is not clear that it is the best pattern to explain what we see. We might ask ourselves if there exists a deeper pattern that provides a better explanation. What Kolmogorov complexity theory teaches is that, at the deepest level, there is no sure way to determine the best pattern. We will simply never know if the pattern that we have found is the best one.

But that makes the search eternally interesting. By definition, something is interesting if it demands more thought. A fact that is obvious and totally understood does not require further thought. The fact that 6 times 7 is 42 is totally comprehensible and uninteresting. It's when we are not certain about ideas that we need to confirm them and think about them. The search for better patterns will always be interesting.

There is an added complexity in the real world. Whereas in the world of strings and computer programs there are no mistakes, in the real world we can, and do, make mistakes. We can easily see if a certain program prints out a string or not. While we might not be able to determine the optimal program to print a certain string, we can determine if the program prints the required string. In contrast, the real world is much more complicated. We can think we recognize a pattern when, in fact, we are mistaken.

Now our understanding of our search for meaning is starting to come together. We abhor randomness and love patterns. We are biologically programmed to find some patterns that explain what we see. But we can never be certain that the pattern we've identified is the right one. Even if we could somehow be assured that we haven't made a mistake, and we are exhibiting a computer-like perfection, there may always still be a deeper truth to unearth.

When we read a newspaper or a history book, we do not desire a simple list of facts or occurrences. We want analysis. We want to be told what patterns or trends are happening. A good writer will tell us what he or she thinks about the facts. How can they be put in order? How will these patterns and trends continue into the cloudy future? We are genetically programmed to try to figure out the patterns so that we are better prepared for the future. The inability to be certain about the pattern—and hence the future—is what makes us constantly search for more, and perhaps better, patterns. We constantly need to impose order on the world around us.

This abhorrence for randomness and the desire for patterns help explain our love of literature, theater, and the cinema. When we read a novel or watch a play, the author or director is presenting us with a sequence of events that has a common theme, pattern, or moral. There would be no literary or movie criticism if literature and movies just described random events and did not have some meaning. The author and director are trying to express some coherent ideas about the universe and the human experience. This is exactly what the audience wants. Human beings are biologically determined to find patterns that explain what they see. It is simple to see such patterns in literature and movies. Most audience members usually go further. Not only do we demand a theme and a moral of the story, we also want a "Hollywood ending." We want the moral of the story to be positive and uplifting. Literature, plays, and the cinema offer us a delightful escape from the usual unintelligible, meaningless chaos that we find in the real world around us.

The search for meaning also defines how we engage with our own lives. While we travel through the seemingly random events in our lives, we are searching for patterns and structure. Life is full of "ups and downs." There are the joys of falling in love, giggling with your child, and feeling a sense of great accomplishment when a hard job is completed. There is also the pain of a crumbling relationship, or the agony of failing at a task after great effort, or the tragedy of the death of a loved one. We try to make sense of all this. We abhor the feeling of total randomness and the idea that we are just following chaotic, habitual laws of physics. We want to know that there is some meaning, purpose, and significance in the world around us. We want a magical story of a life, so we tell ourselves stories.

Sometimes the stories are simply false. Sometimes we lie to ourselves and those around us. And sometimes the patterns we identify are correct. But even if the story is correct, it is not necessarily the best one. We can never know if there is a deeper story that is more exact. As we age and suffer from ennui, we gain certain insights about the universe that we did not see before. We find better patterns. Maybe we get to see things more clearly. Or maybe not. We will never know. But we do know that the search is guaranteed to never end.

Ethics in Statistical Practice and Communication: Five Recommendations

ANDREW GELMAN

"I want to know if it's meant anything," Forlesen said.
"If what I suffered—if it's been worth it."
"No," the little man said. "Yes. No. Yes. Yes. No. Yes. Yes. Maybe."[1]

Statistics and ethics are intertwined, at least in the negative sense, given the famous saying about lies, damn lies, and statistics, and the well-known book, *How to Lie with Statistics* (which, ironically, was written by a journalist with little knowledge of statistics who later accepted thousands of dollars from cigarette companies and told a congressional hearing in 1965 that inferences in the Surgeon General's report on the dangers of smoking were fallacious).

The principle that one should present data as honestly as possible is a fine starting point but does not capture the dynamic nature of science communication: audiences interpret the statistics (and the paragraphs) they read in the context of their understanding of the world and their expectations of the author, who in turn has various goals of exposition and persuasion—and all of this is happening within a competitive publishing environment, in which authors of scientific papers and policy reports have incentives to make dramatic claims.

The result is that scientists are not communicating their work to one another, let alone to general audiences, in terms appropriately geared to enlarging knowledge—they are not doing science properly—and this is one of the recurring threats to the quality of our science communication environment.

Consider this paradox: statistics is the science of uncertainty and variation, but data-based claims in the scientific literature tend to be stated deterministically (e.g. "We have discovered . . . the effect of X on Y is . . . hypothesis H is rejected"). Is statistical communication about exploration and discovery of the unexpected, or is it about making a persuasive, data-based case to back up an argument?

The answer to this question is necessarily each at different times, and sometimes both at the same time. Just as you write in part in order to figure out what you are trying to say, so you do statistics not just to learn from data but also to learn what you can learn from data and to decide how to gather future data to help resolve key uncertainties.

Traditional advice on statistics and ethics focuses on professional integrity, accountability, and responsibility to collaborators and research subjects. All these considerations are important, but when considering ethics, statisticians must also wrestle with fundamental dilemmas regarding the analysis and communication of uncertainty and variation.

In what follows, I make five recommendations for dealing with these dilemmas. These are not intended to be exhaustive, nor do I presume to support them with rigorous quantitative analysis. Rather, they represent recommended directions for progress based on recent experiences.

1. Open Data and Open Methods

Statistical conclusions are data-based, and they can also be, notoriously, dependent on the methods used to analyze the data. An extreme example is the influential paper of Reinhart and Rogoff on the effects of deficit spending, which was used to justify budget-cutting policies.[2] In an infamous mistake, the authors had misaligned columns in an Excel spreadsheet, so their results did not actually follow from their data. This highly consequential error was not detected until years after the article was published and later researchers went to the trouble of replicating the analysis,[3] illustrating how important it is to make data and data-analysis scripts available to others—providing more "eyes on the street," as it were.

There has been much (appropriate) concern about arbitrary decisions in data analysis—"researcher degrees of freedom"[4]—that calls into question the many (most) published p-values in psychology, economics, medicine, etc. But we should also be aware of researcher freedom in data coding, exclusion, and cleaning more generally. Open data and open methods imply a replicable "paper trail" leading from raw data, through processing and statistical analysis, to published conclusions.

Statistics professors promote quantitative measurement, controlled experimentation, careful adjustment in observational studies, and data-based decision making. But in teaching their own classes, they (we) tend to make decisions and inferences based on nonquantitative recall of uncontrolled interventions, just trying things out and seeing what we see—behavior that we would consider laughable and borderline unethical in social or health research.

Are we being unethical in not following our own advice, or in promulgating to others advice we do not ourselves follow? Not necessarily: it is a reasonable position to say that controlled experiments are appropriate in certain medical trials and public interventions, but not in all aspects of our work. However, in that case, we should do a better job of understanding and explaining the conditions under which we do not believe controlled experimentation and statistical analysis to be appropriate.

2. Be Clear about the Information
That Goes into Statistical Procedures

Bayesian inference combines data with prior information, and some Bayesians would argue that it is an ethical requirement to use such methods because otherwise information is being "left on the table" when making decisions. Others take the opposite position and argue that "there are situations where it is very clear that, whatever a scientist or statistician might do privately in looking at data, when they present their information to the public, or government department, or whatever, they should absolutely not use prior information because the prior opinions on some of these prickly issues of public policy can often be highly contentious with different people with strong and very conflicting views."[5]

Both of these extreme views have problems.[6] The recommendation to always use prior information runs into difficulty when this prior information is disputed; in such settings, it makes sense to present unvarnished results. But in many high-stakes settings it is impossible to make any use of data without a model that makes extensive use of prior information. Consider, for example, the reconstruction of historical climate from tree rings, which can only be done in the context of statistical models, which themselves might be contentious. The models relating climate to tree rings are not quite physical models for tree growth and not quite curve fitting, but rather something in between: They are statistical models that are informed by physical considerations. As such, they rely on prior information, even if not in the conventional sense of prior distributions, as discussed by Cox and Mayo in the quote above.[5] The point here is not that Bayes is better (or worse) but that, under any inferential philosophy, we should be able to identify what information is being used in methods.

In some settings, prior information is as strong as or stronger than the data from any given study. For example, Gertler *et al.* reported on an early-childhood intervention performed in an experiment in Jamaica that increased adult earnings (when the children grew up) by an estimated 42%, and the result was statistically significant, thus the data were consistent with effects between roughly 0% and an 80% increase in earnings.[7] But prior knowledge of previous early-childhood interventions suggests that effects of 80%, or even 40%, are implausible. It is fine to present the results from this particular study without reference to any prior information, or to include such information in a non-Bayesian way, as is done in power calculations. But it is not appropriate to offer policy recommendations from this one estimate in isolation. Rather, it is important to understand the implications of the method being used.

3. Create a Culture of Respect for Data

Opacity in data collection, analysis, and reporting is abetted and indeed encouraged by aspects of scholarly research culture. When it comes to data collection, institutional review boards can make it difficult to share one's own data or access others', and when it comes to reporting results, journals favor brevity over completeness. Even in this online age, top journals often aspire to the *Science/Nature* format of three-page articles. Details can appear in online appendixes, but these usually focus not on the specifics of a study but rather on supplementary analyses to buttress the main paper's claims. Published articles typically focus on building a convincing case and giving a sense of certainty, not on making available all the information that would allow outsiders to check and replicate the research.

That said, honesty and transparency are not enough.[8] All the preregistration in the world will not save your study if the data are too remote from the questions of interest. Remoteness can come from mismatch of sample to population, lack of comparability between treatment and control groups, lack of realism of experimental conditions, or, most simply, biased and noisy measurements. For a study to be ethical, it should be informative, which implies serious attention to measurement, design, and data collection. Without good and relevant measurements, protocols such as preregistration, random sampling, and random treatment assignment are empty shells.

As an institutional solution, top journals can publish papers that contain interesting or important data, without the requirement of innovative data analyses or conclusions. In addition to facilitating data availability, this step could also reduce the pressure on researchers to add unnecessary elaborations to their analyses or to hype their conclusions as a way of attaining publication. Public data are important in many fields of study (consider, for example, the U.S. Census, the Panel Study of Income Dynamics, the National Election Study, and various weather and climate databases), so this proposal can be viewed as extending the culture of respect for data and applying it to individual studies.

4. Publication of Criticisms

You do not need to be a philosopher to feel that it is unethical not to admit error or to avoid facing evidence that you have erred. Statistical errors can be technical and hard to notice (and are sometimes even buried within conventional practices, such as taking a statistically significant comparison as strong evidence in favor of a favored hypothesis). Institutions as well as individuals can be averse to admitting error. Indeed, scholarly publishing is often set up to suppress criticism. Journals are notoriously loath to retract articles or publish letters of correction.

For example, a couple of years ago I was pointed to an article in the *American Sociological Review* that suffered from a serious case of selection bias. The article reported that students who paid for their own college education performed better than those who were funded by their parents. But the statistical analysis used to make this claim did not adjust for the fact that self-funded students who were not doing well would be more likely to drop out. Unfortunately, it was not possible to correct this mistake in the journal where it appeared, as the editors judged the correction not to be worthy of publication.

A system of marginalizing criticism creates an incentive for authors to promote dramatic claims, with an upside when published in top journals and little downside if errors are later found. I am sure that the author and editors in this particular case simply made an honest mistake in not catching the selection bias. Nonetheless, the system as a whole gives no clear incentives for the parties involved to be more careful.

Postpublication review outlets, such as PubPeer, and blogs may be changing this equation. This change illustrates the dynamic relation between institutions and ethics that is a theme of the present article. Researchers can do even better by criticizing their own work, as done by Nosek, Spies, and Motyl, who performed an experiment to study "embodiment of political extremism."[9] Their initial finding: "Participants from the political left, right and center ($N = 1979$) completed a perceptual judgment task in which words were presented in different shades of gray. . . . The results were stunning. Moderates perceived the shades of gray more accurately than extremists on the left and right ($p = .01$). Our conclusion: political extremists perceive the world in black-and-white, figuratively and literally."

Before publishing this result, though, the authors decided to collect new data and replicate their study: "We ran 1300 participants, giving us .995 power to detect an effect of the original effect size at $\alpha = .05$."

And then the punch line: "The effect vanished ($p = .59$)."

How did this happen? The original statistically significant result was obtained via a data-dependent analysis procedure. The researchers compared accuracy of perception, but there are many other outcomes they could have looked at: for example, there could have been a correlation with average perceived shade, or an interaction with age, sex, or various other logical moderators, or an effect just for Democrats or just for Republicans, and so forth. The replication, with its prechosen comparison, was not subject to this selection effect.

Nosek *et al.* discuss how to reform the scientific publication system to provide incentives for this self-critical behavior. But in the meantime, you can do it yourself—just as they did!

More generally, you can make self-criticism part of your general practice by enabling others' criticisms of your work, via open data, clarity in assumptions, and the other steps listed here. Joining with others to criticize your own practices should strengthen your work. These recommendations on facilitating criticisms are consistent with the American Statistical Association's recent ethical guidelines, which call for prompt correction of errors and appropriate dissemination of the correction (bit.ly/2ML136N).

5. Respect the Limitations of Statistics

Many fields of empirical research have become notorious for claims published in serious journals that make little sense (for example, the claim that people react differently to hurricanes with male and female names[10] or the claim that women have dramatically different political preferences at different times of the month, or the claim that the subliminal image of a smiley face has large effects on attitudes on immigration policy[11]) but which are easily understood as the inevitable product of explicit or implicit searches for statistical significance with flexible hypotheses that are rich in researcher degrees of freedom.[4] Unsurprisingly (given this statistical perspective), several high-profile research papers in social psychology have failed to replicate, for example, the well-publicized claim in "embodied cognition" that college students walk more slowly after being subtly primed by being exposed to elderly-related words.[12]

Just to be clear, the above claims seem to many people (including the current author) to be silly, but they are not impossible—at least in a qualitative sense. For example, the literature on public opinion makes it highly implausible that women were experiencing during their monthly cycles a 20% swing in probability of supporting Barack Obama for president, as claimed by Durante, Arsena, and Griskevicius.[13] It is, however, possible that there is a tiny effect, essentially undetectable in the study in question, given the precision of measurement of the relevant variables.[14]

The error in that paper (and in the hurricanes paper and the others mentioned above) is that the data do not provide strong evidence for the authors' claims. These papers, and the system by which they are published and publicized, represent a failure in science communication in that they place an impossible burden on statistical data collection and analysis.

Moving away from Systematic Overconfidence

In statistics, we use mathematical analysis and stochastic simulation to evaluate the properties of proposed designs and data analyses.

Recommendations for ethics are qualitative and cannot be evaluated in such formal ways. Nonetheless, there is value in the recommendations made in this paper and in emphasizing the links between ethical principles and the general statistical concepts of variation and uncertainty.

So far, this is just a story of statistical confusion, perhaps abetted by incentives toward reporting dramatic claims on weak evidence. The ethics come in if we think of this entire journal publication system as a sort of machine for laundering uncertainty: Researchers start with junk data (for example, poorly thought-out experiments on college students, or surveys of online Amazon Mechanical Turk participants) and then work with the data, straining out the null results and reporting what is statistically significant, in a process analogous to the notorious mortgage lenders of the mid-2000s, who created high-value "tranches" out of subprime loans. The loan crisis precipitated an economic recession, and I doubt that the replication crisis will trigger such a crash in science. But I see a crucial similarity in that technical methods (e.g., structured finance for mortgages or statistical significance for scientific research) were being used to create value out of thin air.

In their article, "The AAA tranche of subprime science," Loken and Gelman concluded

> When we as statisticians see researchers making strong conclusions based on analyses affected by selection bias, multiple comparisons, and other well-known threats to statistical validity, our first inclination might be to throw up our hands and feel we have not been good teachers, that we have not done a good enough job conveying our key principles to the scientific community.
>
> But maybe we should consider another, less comforting possibility, which is that our fundamental values have been conveyed all too well and the message we have been sending—all too successfully—is that statistics is a form of modern alchemy, transforming the uncertainty and variation of the laboratory and field measurements into clean scientific conclusions that can be taken as truth . . .
>
> We have to make personal and political decisions about health care, the environment, and economics—to name only a few areas—in the face of uncertainty and variation. It's exactly because we have a tendency to think more categorically about things

as being true or false, there or not there, that we need statistics. Quantitative research is our central tool for understanding variance and uncertainty and should not be used as a way to overstate confidence.[15]

Ethics is, in this way, central to statistics and public policy. We use statistics to measure uncertainty and variation, but all too often we sell our methods as a sort of alchemy that will transform these into certainty. The first step to not fooling others is to not fool ourselves.

Acknowledgments

The author wishes to thank two reviewers for helpful comments and the U.S. Office of Naval Research Grant N00014-15-1-2541 and Defense Advanced Research Projects Agency Grant D17AC00001 for partial support of this work.

References

1. Wolfe, G. (1974) Forlesen. In D. Knight (ed.), *Orbit 14*. New York: Harper & Row.
2. Reinhart, C. M. and Rogoff, K. S. (2010) Growth in a time of debt. *American Economic Review, 100*, 573–578.
3. Herndon, T., Ash, M. and Pollin, R. (2014) Does high public debt consistently stifle economic growth? A critique of Reinhart and Rogoff. *Cambridge Journal of Economics, 38*, 257–279.
4. Simmons, J. P., Nelson, L. D. and Simonsohn, U. (2011) False-positive psychology: Undisclosed flexibility in data collection and analysis allow presenting anything as significant. *Psychological Science, 22*, 1359–1366.
5. Cox, D. R. and Mayo, D. (2011) A statistical scientist meets a philosopher of science: A conversation. *Rationality, Markets and Morals, 2*, 103–114.
6. Gelman, A. (2012) Ethics and the statistical use of prior information. *Chance, 25*(4), 52–54.
7. Gertler, P., Heckman, J., Pinto, R., Zanolini, A., Vermeerch, C., Walker, S., Chang, S. M. and Grantham-McGregor, S. (2013) Labor market returns to early childhood stimulation: A 20-year followup to an experimental intervention in Jamaica. IRLE Working Paper No. 142-13. Institute for Research on Labor and Employment, Berkeley, CA. http://irle.berkeley.edu/ workingpapers/142-13.pdf.
8. Gelman, A. (2017) Honesty and transparency are not enough. *Chance, 30*(1), 37–39.
9. Nosek, B. A., Spies, J. R. and Motyl, M. (2012) Scientific utopia II. Restructuring incentives and practices to promote truth over publishability. *Perspectives on Psychological Science, 7*, 615–631.
10. Malter, D. (2014) Female hurricanes are not deadlier than male hurricanes. *Proceedings of the National Academy of Sciences of the USA, 111*, E3496.
11. Gelman, A. (2015) Disagreements about the strength of evidence. *Chance, 28*, 55–59.

12. Doyen, S., Klein, O., Pichon, C. L. and Cleeremans, A. (2012) Behavioral priming: It's all in the mind, but whose mind? *PLoS ONE, 7,* e29081.

13. Durante, K. M., Arsena, A. R. and Griskevicius, V. (2013) The fluctuating female vote: Politics, religion, and the ovulatory cycle. *Psychological Science, 24,* 1007–1016.

14. Gelman, A. (2015) The connection between varying treatment effects and the crisis of unreplicable research: A Bayesian perspective. *Journal of Management, 41,* 632–643.

15. Loken, E., and Gelman, A. (2014) The AAA tranche of subprime science. *Chance, 27*(1), 51–56.

The Fields Medal Should Return to Its Roots

Michael J. Barany

Like Olympic medals and World Cup trophies, the best-known prizes in mathematics come around only every four years. Already, math departments around the world are buzzing with speculation: 2018 is a Fields Medal year.

While looking forward to this year's announcement, I've been looking backward with an even keener interest. In long-overlooked archives, I've found details of turning points in the medal's past that, in my view, hold lessons for those deliberating whom to recognize in August at the 2018 International Congress of Mathematicians in Rio de Janeiro in Brazil, and beyond.

Since the late 1960s, the Fields Medal has been popularly compared to the Nobel prize, which has no category for mathematics.[1] In fact, the two are different in their procedures, criteria, remuneration, and much else. Notably, the Nobel is typically given to senior figures, often decades after the contribution being honored. By contrast, Fields medalists are at an age at which, in most sciences, a promising career would just be taking off.

This idea of giving a top prize to rising stars who—by brilliance, luck, and circumstance—happen to have made a major mark when relatively young is an accident of history. It is not a reflection of any special connection between math and youth—a myth unsupported by the data.[2,3] As some mathematicians have long recognized,[4] this accident has been to the detriment of mathematics. It reinforces biases within the discipline and in the public's attitudes about mathematicians' work, career pathways, and intellectual and social values. All 56 winners so far have been phenomenal mathematicians, but such biases have contributed to 55 of them being male, most coming from the United States and Europe, and most working on a collection of research topics that are arguably unrepresentative of the discipline as a whole.

When it began in the 1930s, the Fields Medal had different goals. It was rooted more in smoothing over international conflict than in celebrating outstanding scholars. In fact, early committees deliberately avoided trying to identify the best young mathematicians and sought to promote relatively unrecognized individuals. As I demonstrate here, they used the medal to shape their discipline's future, not just to judge its past and present.

As the mathematics profession grew and spread, the number of mathematicians and the variety of their settings made it harder to agree on who met the vague standard of being promising, but not a star. In 1966, the Fields Medal committee opted for the current compromise of considering all mathematicians under the age of 40. Instead of celebrity being a disqualification, it became almost a prerequisite.

I think that the Fields Medal should return to its roots. Advanced mathematics shapes our world in more ways than ever, the discipline is larger and more diverse, and its demographic issues and institutional challenges are more urgent. The Fields Medal plays a big part in defining what and who matters in mathematics.

The committee should leverage this role by awarding medals on the basis of what mathematics can and should be, not just what happens to rise fastest and shine brightest under entrenched norms and structures. By challenging themselves to ask every four years which unrecognized mathematics and mathematicians deserve a spotlight, the prize givers could assume a more active responsibility for their discipline's future.

Born of Conflict

The Fields Medal emerged from a time of deep conflict in international mathematics that shaped the conceptions of its purpose. Its chief proponent was John Charles Fields, a Canadian mathematician who spent his early career in a fin de siècle European mathematical community that was just beginning to conceive of the field as an international endeavor.[5]

The first International Congress of Mathematicians (ICM) took place in 1897 in Zurich, Switzerland, followed by ICMs in Paris in 1900, Heidelberg, Germany, in 1904, Rome in 1908, and Cambridge,

U.K., in 1912. The First World War derailed plans for a 1916 ICM in Stockholm and threw mathematicians into turmoil.

When the dust settled, aggrieved researchers from France and Belgium took the reins and insisted that Germans and their wartime allies had no part in new international endeavors, congresses or otherwise. They planned the first postwar meeting for 1920 in Strasbourg, a city just repatriated to France after half a century of German rule.

In Strasbourg, the U.S. delegation won the right to host the next ICM, but when its members returned home to start fundraising, they found that the rule of German exclusion dissuaded many potential supporters. Fields took the chance to bring the ICM to Canada instead. In terms of international participation, the 1924 Toronto congress was disastrous, but it finished with a modest financial surplus. The idea for an international medal emerged in the organizers' discussions, years later, over what to do with these leftover funds.

Fields forced the issue from his deathbed in 1932, endowing two medals to be awarded at each ICM. The 1932 ICM in Zurich appointed a committee to select the 1936 medalists but left no instructions as to how the group should proceed. Instead, early committees were guided by a memorandum that Fields wrote shortly before his death, titled "International Medals for Outstanding Discoveries in Mathematics."

Most of the memorandum is procedural: how to handle the funds, appoint a committee, communicate its decision, design the medal, and so on. In fact, Fields wrote, the committee "should be left as free as possible" to decide winners. To minimize national rivalry, Fields stipulated that the medal should not be named after any person or place, and he never intended for it to be named after himself. His most famous instruction, later used to justify an age limit, was that the awards should be both "in recognition of work already done" and "an encouragement for further achievement." But in context, this instruction had a different purpose: "to avoid invidious comparisons" among factious national groups over who deserved to win.

The first medals were awarded in 1936, to mathematicians Lars Ahlfors from Finland and Jesse Douglas from the United States. The Second World War delayed the next medals until 1950. They have been given every four years since.

Blood and Tears

The Fields Medal selection process is supposed to be secret, but mathematicians are human. They gossip and, luckily for historians, occasionally neglect to guard confidential documents. Especially for the early years of the Fields Medal, before the International Mathematical Union became more formally involved in the process, such ephemera may well be the only extant records.

One of the 1936 medalists, Ahlfors, served on the committee to select the 1950 winners. His copy of the committee's correspondence made its way into a mass of documents connected with the 1950 ICM, largely hosted by Ahlfors's department at Harvard University in Cambridge, Massachusetts; these are now in the university's archives.

The 1950 Fields Medal committee had broad international membership. Its chair, Harald Bohr (younger brother of the physicist Niels), was based in Denmark. Other members hailed from Cambridge, U.K.; Princeton, New Jersey; Paris; Warsaw; and Bombay. They communicated mostly through letters sent to Bohr, who summarized the key points in letters he sent back. The committee conducted most of these exchanges in the second half of 1949, agreeing on the two winners that December.

The letters suggest that Bohr entered the process with a strong opinion about who should win one of the medals: the French mathematician Laurent Schwartz, who had blown Bohr away with an exciting new theory at a 1947 conference.[6] The Second World War meant that Schwartz's career had got off to an especially rocky start: he was Jewish and a Trotskyist, and he spent part of the French Vichy regime in hiding using a false name. His long-awaited textbook had still not appeared by the end of 1949, and there were few major new results to show.

Bohr saw in Schwartz a charismatic leader of mathematics who could offer new connections between pure and applied fields. Schwartz's theory did not have quite the revolutionary effects Bohr predicted, but, by promoting it with a Fields Medal, Bohr made a decisive intervention oriented toward his discipline's future.

The best way to ensure that Schwartz won, Bohr determined, was to ally with Marston Morse of the Institute for Advanced Study in Princeton, who in turn was promoting his Norwegian colleague, Atle Selberg.

The path to convincing the rest of the committee was not straightforward, and their debates reveal a great deal about how the members thought about the Fields Medal.

Committee members started talking about criteria such as age and fields of study, even before suggesting nominees. Most thought that focusing on specific branches of mathematics was inadvisable. They entertained a range of potential age considerations, from an upper limit of 30 to a general principle that nominees should have made their mark in mathematics some time since the previous ICM in 1936. Bohr cryptically suggested that a cutoff of 42 "would be a rather natural limit of age."

By the time the first set of nominees was in, Bohr's cutoff seemed a lot less arbitrary. It became clear that the leading threat to Bohr's designs for Schwartz was another French mathematician, André Weil, who turned 43 in May 1949. Everyone, Bohr and Morse included, agreed that Weil was the more accomplished mathematician. But Bohr used the question of age to try to ensure that he didn't win.

As chair, Bohr had some control over the narrative, frequently alluding to members' views that "young" mathematicians should be favored while framing Schwartz as the prime example of youth. He asserted that Weil was already "too generally recognized" and drew attention to Ahlfors's contention that to give a medal to Weil would be "maybe even disastrous" because "it would make the impression that the Committee has tried to designate the greatest mathematical genius."

Their primary objective was to avoid international conflict and invidious comparisons. If they could deny having tried to select the best, they couldn't be accused of having snubbed someone better.

But Weil wouldn't go away. Committee member Damodar Kosambi thought it would be "ridiculous" to deny him a medal—a comment Bohr gossiped about to a Danish colleague but did not share with the committee. Member William Hodge worried "whether we might be shirking our duty" if Weil did not win. Even Ahlfors argued that they should expand the award to four recipients so that they could include Weil. Bohr wrote again to his Danish confidant that "it will require blood and tears" to seal the deal for Schwartz and Selberg.

Bohr prevailed by cutting the debate short. He argued that choosing Weil would open a floodgate to considering prominent older mathematicians, and asked for an up or down vote on the pair of Schwartz and Selberg. Finally, at the awards ceremony at the 1950 ICM, Bohr praised Schwartz for being recognized and eagerly followed by a younger generation of mathematicians—the very attributes he had used to exclude Weil.

Further Encouragement

Another file from the Harvard archives shows that the 1950 deliberations reflected broader attitudes toward the medal, not just one zealous chair's tactics. Harvard mathematician Oscar Zariski kept a selection of letters from his service on the 1958 committee in his private collection.

Zariski's committee was chaired by mathematician Heinz Hopf of the Swiss Federal Institute of Technology in Zurich. Its first round of nominations produced 38 names. Friedrich Hirzebruch was the clear favorite, proposed by five of the committee members.

Hopf began by crossing off the list the two oldest nominees, Lars Gårding and Lipman Bers. His next move proved that it was not age per se that was the real disqualifying factor, but prior recognition: he ruled out Hirzebruch and one other who, having recently taken up professorships at prestigious institutions, "did not need further encouragement." Nobody on the committee seems to have batted an eyelid.

Of those remaining, the committee agreed that Alexander Grothendieck was the most talented, but few of his results were published, and they considered him a shoo-in for 1962. John Nash, born in the same year as Grothendieck (1928), came third in the final ballot. Although the 1958 short list also included Olga Ladyzhenskaya and Harish-Chandra, it would take until 2014 for the Fields Medals to go to a woman (Maryam Mirzakhani) or a mathematician of Indian descent (Manjul Bhargava). Ultimately, the 1958 awards went to Klaus Roth and René Thom, both of whom the committee considered promising but not too accomplished—unlikely to provoke invidious comparisons.

A Sweeping Expedient

By 1966, the adjudication of which young mathematicians were good but not too good was growing imponderable. That year, committee chair Georges de Rham adopted a firm age limit of 40, the smallest round number that covered the ages of all the previous Fields recipients.

Suddenly, mathematicians who would previously have been considered too accomplished were eligible. Grothendieck, presumably ruled out as too well-known in 1962, was offered the medal in 1966 but boycotted its presentation for political reasons.

The 1966 cohort contained another politically active mathematician, Stephen Smale. He went to accept his medal in Moscow rather than testify before the U.S. House Un-American Activities Committee about his activism against the Vietnam War. Colleagues' efforts to defend the move were repeated across major media outlets, and the "Nobel prize of mathematics" moniker was born.

This coincidence—comparing the Fields Medal to a higher-profile prize at the same time that a rule change allowed the medalists to be much more advanced—had a lasting impact in mathematics and on the award's public image. It radically rewrote the medal's purpose, divorcing it from the original goal of international reconciliation and embracing precisely the kinds of judgement Fields thought would only reinforce rivalry.

Any method of singling out a handful of honorees from a vast discipline has shortcomings and controversies. Social and structural circumstances affect who has the opportunity to advance in the discipline at all stages, from primary school to the professoriate. Selection committees themselves need to be diverse and attuned to the complex values and roles of mathematics in society.

But however flawed the processes were before 1966, they forced a committee of elite mathematicians to think hard about their discipline's future. The committees used the medal as a redistributive tool, to give a boost to those whom they felt did not already have every advantage but were doing important work nonetheless.

Our current understanding of the social effect of mathematics and of barriers to diversity within it is decidedly different to that of mathematicians in the mid-twentieth century. If committees today were given the same license to define the award that early committees enjoyed, they could focus on mathematicians who have backgrounds and identities that are underrepresented in the discipline's elite. They could promote areas of study on the basis of the good they do in the world, beyond just the difficult theorems they produce.

In my view, the medal's history is an invitation for mathematicians today to think creatively about the future, and about what they could say collectively with their most famous award.

Notes

1. Barany, M. J. *Not. Am. Math. Soc.* **62,** 15–20 (2015).
2. Stern, N. *Soc. Stud. Sci.* **8,** 127–140 (1978).
3. Hersh, R. & John-Steiner, V. *Loving + Hating Mathematics: Challenging the Myths of Mathematical Life* 251–272 (Princeton Univ. Press, 2011).
4. Henrion, C. *AWM Newsletter* **25** (6), 12–16 (1995).
5. Riehm, E. M. & Hoffman, F. *Turbulent Times in Mathematics: The Life of J.C. Fields and the History of the Fields Medal* (American Mathematical Society & Fields Institute, 2011).
6. Barany, M. J., Paumier, A.-S. & Lützen, J. *Hist. Math.* **44,** 367–394 (2017).

The Erdős Paradox

MELVYN B. NATHANSON

Prologue

The great Hungarian mathematician Paul Erdős was born in Budapest on March 26, 1913. He died alone in a hospital room in Warsaw, Poland, on Friday afternoon, September 20, 1996. It was sad and ironic that he was alone because he probably had more friends in more places than any mathematician in the world. He was in Warsaw for a conference. Vera Sós had also been there but had gone to Budapest on Thursday and intended to return on Saturday with András Sárközy to travel with Paul to a number theory meeting in Vilnius. On Thursday night, Erdős felt ill and called the desk in his hotel. He was having a heart attack and was taken to a hospital, where he died about 12 hours later. No one knew he was in the hospital. When Paul did not appear at the meeting on Friday morning, one of the Polish mathematicians called the hotel. He did not get through, and no one tried to telephone the hotel again for several hours. By the time it was learned that Paul was in the hospital, he was dead.

Vera was informed by telephone on Friday afternoon that Paul had died. She returned to Warsaw on Saturday. It was decided that Paul should be cremated. This was contrary to Jewish law, but Paul was not an observant Jew and it is not known what he would have wanted. Nor was he buried promptly in accordance with Jewish tradition. Instead, four weeks later, on October 18, there was a secular funeral service in Budapest, and his ashes were buried in the Jewish cemetery in Budapest.

Erdős strongly identified with Hungary and with Judaism. He was not religious, but he visited Israel often and established a mathematics prize and a postdoctoral fellowship there. He also established a prize and a lectureship in Hungary. He told me that he was happy whenever someone proved a beautiful theorem, but that he was especially happy if the person who proved the theorem was Hungarian or Jewish.

Mathematicians from the United States, Israel, and many European countries traveled to Hungary to attend Erdős's funeral. The following day, a conference entitled "Paul Erdős and his Mathematics" took place at the Hungarian Academy of Sciences in Budapest, and mathematicians who were present for the funeral were asked to lecture on different parts of Erdős's work. I was asked to chair one of the sessions and to begin with some personal remarks about my relationship with Erdős and his life and style.

This paper is in two parts. The first is the verbatim text of my remarks at the Erdős memorial conference in Budapest on October 19, 1996. A few months after the funeral and conference, I returned to Europe to lecture in Germany. At Bielefeld, someone told me that my eulogy had generated controversy, and indeed, I heard the same report a few weeks later when I was back in the United States. Eighteen years later, on the 100th anniversary of his birth, it is fitting to reconsider Erdős's life and work.

1. Eulogy, Delivered in Budapest on October 19, 1996

I knew Erdős for 25 years, half my life, but still not very long compared to many people in this room. His memory was much better than mine; he often reminded me that we proved the theorems in our first paper in 1972 in a car as we drove back to Southern Illinois University in Carbondale after a meeting of the Illinois Number Theory Conference in Normal, Illinois. He visited me often in Carbondale and even more often after I moved to New Jersey. He would frequently leave his winter coat in my house when he left for Europe in the spring and retrieve it when he returned in the fall. I still have a carton of his belongings in my attic. My children Becky and Alex, who are five and seven years old, would ask, "When is Paul coming to visit again?" They liked his silly tricks for kids, like dropping a coin and catching it before it hit the floor. He was tolerant of the dietary rules in my house, which meant, for example, no milk in his espresso if we had just eaten meat.

He was tough. "No illegal thinking," he would say when we were working together. This meant no thinking about mathematical problems other than the ones we were working on at that time. In other words, he knew how to enforce party discipline.

Erdős loved to discuss politics, especially Sam and Joe, which, in his idiosyncratic language, meant the USA (Uncle Sam) and the Soviet Union (Joseph Stalin). His politics seemed to me to be the politics of the 30s, much to the left of my own. He embraced a kind of naive and altruistic socialism that I associate with idealistic intellectuals of his generation. He never wanted to believe what I told him about the Soviet Union as an "evil empire." I think he was genuinely saddened by the fact that the demise of communism in the Soviet Union meant the failure of certain dreams and principles that were important to him.

Erdős's cultural interests were narrowly focused. When he was in my house, he always wanted to hear "noise" (that is, music), especially Bach. He loved to quote Hungarian poetry (in translation). I assume that when he was young he read literature (he was amazed that Anatole France is a forgotten literary figure today), but I don't think he read much anymore.

I subscribe to many political journals. When he came to my house, he would look for the latest issue of *Foreign Affairs* but usually disagreed with the contents. Not long ago, an American historian at Pacific Lutheran University published a book entitled *Ordinary Men*,[1] a study of how large numbers of "ordinary Germans," not just a few SS, actively and willingly participated in the murder of Jews. He found the book on my desk and read it, but did not believe or did not want to believe it could be true, because it conflicted with his belief in the natural goodness of ordinary men.

He had absolutely no interest in the visual arts. My wife was a curator at the Museum of Modern Art in New York, and we went with her one day to the museum. It has the finest collection of modern art in the world, but Paul was bored. After a few minutes, he went out to the sculpture garden and started, as usual, to prove and conjecture.

Paul's mathematics was like his politics. He learned mathematics in the 1930s in Hungary and England, and England at that time was a kind of mathematical backwater. For the rest of his life, he concentrated on the fields that he had learned as a boy: elementary and analytic number theory, at the level of Landau, graph theory, set theory, probability theory, and classical analysis. In these fields, he was an absolute master, a virtuoso.

At the same time, it is extraordinary to think of the parts of mathematics he never learned. Much of contemporary number theory, for example. In retrospect, probably the greatest number theorist of the 1930s was Hecke, but Erdős knew nothing about his work and cared less. Hardy and Littlewood dominated British number theory when Erdős lived in England, but I doubt they understood Hecke.

There is an essay by Irving Segal[2] in the current issue of the *Bulletin of the American Mathematical Society*. He tells the story of the visit of another great Hungarian mathematician, John von Neumann, to Cambridge in the 1930s. After his lecture, Hardy remarked, "Obviously a very intelligent young man. But was that *mathematics*?"

A few months ago, on his last visit to New Jersey, I was telling Erdős something about p-adic analysis. Erdős was not interested. "You know," he said about the p-adic numbers, "they don't really exist."

Paul never learned algebraic number theory. He was offended—actually, he was furious—when André Weil wrote that analytic number theory is good mathematics, but analysis, not number theory.[3] Paul's "tit-for-tat" response was that André Weil did good mathematics, but it was algebra, not number theory. I think Paul was a bit shocked that a problem he did consider number theory, Fermat's Last Theorem, was solved using ideas and methods of Weil and other very sophisticated mathematicians.

It is idle to speculate about how great a mathematician Erdős was, as if one could put together a list of the top 10 or top 100 mathematicians of our century. His interests were broad, his conjectures, problems, and results profound, and his humanity extraordinary.

He was the Bob Hope of mathematics, a kind of vaudeville performer who told the same jokes and the same stories a thousand times. When he was scheduled to give yet another talk, no matter how tired he was, as soon as he was introduced to the audience, the adrenaline (or maybe amphetamine) would release into his system, and he would bound onto the stage, full of energy, and do his routine for the 1001st time.

If he were here today, he would be sitting in the first row, half asleep, happy to be in the presence of so many colleagues, collaborators, and friends.

יתגדל ויתקדש שמה רבה
יהי זכרונו לעולם

May his memory be with us forever.[4]

2. Reconsideration

My brief talk at the Erdős conference was not intended for publication. Someone asked me for a copy, and it subsequently spread via e-mail. Many people who heard me in Budapest or who later read my eulogy told me that it helped them remember Paul as a human being, but others clearly disliked what I said. I confess that I still don't know what disturbed them so deeply. It has less to do with Erdős, I think, than with the status of Hungarian mathematics in the scientific world.[5]

Everyone understands that Erdős was an extraordinary human being and a great mathematician who made major contributions to many parts of mathematics. He was a central figure in the creation of new fields, such as probabilistic number theory and random graphs. This part of the story is trivial.

It is also true, understood by almost everyone, and not controversial, that Erdős did not work in and never learned the central core of twentieth-century mathematics. It is amazing to me how great were Erdős's contributions to mathematics and yet how little he knew. He never learned, for example, the great discoveries in number theory that were made at the beginning of the twentieth century. These include, for example, Weil's work on Diophantine equations, Artin's class field theory, and Hecke's monumental contributions to modular forms and analytic number theory. Erdős apparently knew nothing about Lie groups, Riemannian manifolds, algebraic geometry, algebraic topology, global analysis, or the deep ocean of mathematics connected with quantum mechanics and relativity theory. These subjects, already intensely investigated in the 1930s, were at the heart of twentieth-century mathematics. How could a great mathematician not want to study these things?[6] This is the first Erdős paradox.

In the case of the Indian mathematician Ramanujan, whose knowledge was also deep but narrow, there is a discussion in the literature about the possible sources of his mathematical education. The explanation of Hardy[7] and others is that the only serious book that was accessible to Ramanujan in India was Carr's *A Synopsis of Elementary Results in Pure and Applied Mathematics* and that Ramanujan lacked a broad mathematical culture because he did not have access to books and journals in India. But Hungary was not India; there were libraries, books, and journals in Budapest, and in other places where Erdős lived in the 1930s and 1940s.

For the past half-century, "Hungarian mathematics" has been a term of art to describe the kind of mathematics that Erdős did.[8] It includes combinatorics, graph theory, combinatorial set theory, and elementary and combinatorial number theory. Not all Hungarians do this kind of mathematics, of course, and many non-Hungarians do Hungarian mathematics. It happens that combinatorial reasoning is central to theoretical computer science and that "Hungarian mathematics" commands vast respect in the computer science world. It is also true, however, that for many years combinatorics did not have the highest reputation among mathematicians in the ruling subset of the research community, exactly because combinatorics was concerned largely with questions that they believed (incorrectly) were not central to twentieth-century mathematics.[9]

In a volume in honor of Erdős's 70th birthday, Ernst Straus wrote, "In our century, in which mathematics is so strongly dominated by 'theory constructors' [Erdős] has remained the prince of problem solvers and the absolute monarch of problem posers."[10] I disagree. There is, as Gel'fand often said, only one mathematics. There is no separation of mathematics into "theory" and "problems." But there is an interesting lurking issue.

In his lifetime, did Erdős get the recognition he deserved? Even though Erdős received almost every honor that can be given to a mathematician, some of his friends believe that he was still insufficiently appreciated, and they are bitter on his behalf.

He was awarded a Wolf Prize and a Cole Prize, but he did not get a Fields Medal or a permanent professorship at the Institute for Advanced Study. He traveled from one university to another across the United States and was never without an invitation to lecture somewhere, but his mathematics was not highly regarded by the power brokers of mathematics. To them, his methods were insufficiently abstruse and obscure; they did not require complicated machinery. Paul invented diabolically clever arguments from arithmetic, combinatorics, and probability to solve problems. But the technique was too simple, too elementary. It was suspicious. The work could not be "deep."

None of this seemed to matter to Erdős, who was content to prove and conjecture and publish more than 1,500 papers.

Not because of politicking, but because of computer science and because his mathematics was always beautiful, in the past decade the reputation of Erdős and the respect paid to discrete mathematics have increased exponentially. The *Annals of Mathematics* now publishes papers in combinatorics, and the most active seminar at the Institute for Advanced Study is in discrete mathematics and theoretical computer science. Fields Medals are awarded to mathematicians who solve Erdős-type problems. Science has changed.

In 1988, Alexander Grothendieck was awarded the Crafoord Prize of the Swedish Academy of Sciences. In the letter to the Swedish Academy in which he declined the prize, he wrote, "Je suis persuadé que la seule épreuve décisive pour la fécondité d'idées ou d'une vision nouvelles est celle du temps. La fécondité se reconnait par la progéniture, et non par les honneurs."[11]

Time has proved the fertility and richness of Erdős's work. The second Erdős paradox is that his methods and results, considered marginal in the twentieth century, have become central in twenty-first-century mathematics.

May his memory be with us forever.

Notes

1. Christopher R. Browning, *Ordinary Men*, HarperCollins Publishers, New York, 1992.

2. Irving Segal. "*Noncommutative Geometry* by Alain Connes (book review)," *Bull. Amer. Math. Soc.* 33 (1996), 459–465.

3. Weil wrote, ". . . there is a subject in mathematics (it's a perfectly good and valid subject and it's perfectly good and valid mathematics) which is called Analytic Number Theory. . . . I would classify it under analysis . . . ." (*Œuvres Scientifiques Collected Papers*, Springer-Verlag, New York, 1979, Volume III, p. 280).

4. I ended my eulogy with a sentence in Aramaic and a sentence in Hebrew. The first is the first line of the Kaddish, the Jewish prayer for the dead. Immediately following the second sentence is its English translation.

5. cf. L. Babai, "In and out of Hungary: Paul Erdős, his friends, and times," in *Combinatorics, Paul Erdős Is Eighty (Vol. 2), Keszthely (Hungary) 1993*, Bolyai Society Mathematical Studies, Budapest, 1996, pp. 7–95.

6. This notion suggests the fundamental question: How much, or how little, must one know in order to do great mathematics?

7. "It was a book of a very different kind, Carr's *Synopsis*, which first aroused Ramanujan's full powers," according to G. H. Hardy, in his book *Ramanujan*, Chelsea Publishing. New York, 1959, p. 2.

8. For example. Joel Spencer said, "I felt . . . I was working on 'Hungarian mathematics,'" quoted in Babai, *op. cit.*

9. For example, S. Mac Lane criticized "emphasizing too much of a Hungarian view of mathematics," in: "The health of mathematics," *Math.Intelligencer* 5 (1983), 53–55.

10. E. G. Straus, "Paul Erdős at 70," *Combinatorica* 3 (1983), 245–246. Tim Gowers revisited this notion in his essay, "The two cultures of mathematics," published in *Mathematics: Frontiers and Perspectives*, American Mathematical Society, 2000.

11. "I believe that time gives the only definite proof of the fertility of new ideas or a new vision. We recognize fertility by its offspring, and not by honors."

Contributors

Jeremy Avigad is a professor of philosophy and mathematical sciences at Carnegie Mellon University, with an MA in Mathematics from Harvard (1989) and a Ph.D. in mathematics from the University of California, Berkeley (1995). He was trained as a mathematical logician in the tradition of Hilbert's *Beweistheorie* and is generally interested in applications of logic to mathematics, philosophy, and computer science. He has worked on philosophical theories of mathematical understanding and has also done research in the history of mathematics, with a focus on nineteenth-century mathematics in particular. Finally, he has a long-standing interest in formal methods in mathematics and interactive theorem proving. He has coauthored formalizations of the prime number theorem and the central limit theorem in Isabelle, and he contributed to the landmark formalization of the Feit-Thompson theorem in Coq. He is currently involved in the Formal Abstracts project and the development of the mathematical library for Lean, a new interactive theorem prover.

Michael J. Barany (http://orcid.org/0000-0002-4067-5112) studies the institutional, political, social, conceptual, and material dimensions of abstract knowledge in modern societies and is preparing a book on the globalization of the mathematics profession in the twentieth century. He is currently a lecturer in the history of science at the University of Edinburgh, after completing his Ph.D. at Princeton University (2016) and holding postdoctoral fellowships at Dartmouth College (2016–2018) and Northwestern University (2018). This is his fourth appearance in the *Best Writing on Mathematics* series. You can read more of his work at http://mbarany.com and follow him on Twitter at @mbarany.

Tiziana Bascelli holds a Ph.D. in theoretical philosophy, a BA in mathematics, and a BA in philosophy. She obtained all three degrees at the University of Padua, Italy. She has edited and coedited a number of books and published several scientific publications on the history of science. Her field of research is interdisciplinary. She is mostly interested in the shaping of modern science through mathematics, examined from the point of view of history and epistemology, with an integrated but also constructivist approach. In particular, she studies the transformation of natural philosophy into modern science by focusing on key concepts related to motion, matter, and force.

Piotr Błaszczyk (https://orcid.org/0000-0002-3501-348) is a professor at the Institute of Mathematics, Pedagogical University (Kraków, Poland). He obtained degrees in mathematics (1986), philosophy (1994), and a Ph.D. in philosophy (2002) from Jagiellonian University (Kraków). He authored "Philosophical Analysis of Richard Dedekind's memoir *Stetigkeit und irrationale Zahlen*" (2007, Habilitationsschrift), and coauthored "Euclid, Elements, Books V–VI. Translation and Commentary" (2013), "Descartes, Geometry. Translation and Commentary" (2015), and "Descartes, Dioptrics. Translation and Commentary" (2018). His research interest is the idea of continuum and continuity from Euclid to modern times. His recent papers on nonstandard analysis include "Calculus without the Concept of Limit" (coauthored with J. Major) and "A Purely Algebraic Proof of the Fundamental Theorem of Algebra."

Paul J. Campbell (MR Author ID: 44685, https://orcid.org/0000-0002-0556-5540) graduated summa cum laude from the University of Dayton and received an M.S. in algebra and a Ph.D. in mathematical logic from Cornell University, where he was a Woodrow Wilson, Danforth, and NSF Graduate Fellow and a teaching assistant in statistics under Jack Kiefer. He taught at St. Olaf College, has been a frequent guest professor at the University of Augsburg in Germany, and recently retired from Beloit College. He has been the reviews editor for *Mathematics Magazine* since 1977 and editor of *The UMAP Journal of Undergraduate Mathematics and Its Applications* since 1984. He is a coauthor of *For All Practical Purposes* (11th ed., W.H. Freeman, forthcoming), a textbook of applications-oriented introductory collegiate mathematics that has been used by more than 1 million students. He is interested in everything.

Toby S. Cubitt's research in quantum information theory straddles theoretical physics, mathematics, computer science, and much of Europe. He originally hails from Luxembourg. His undergraduate degree from the University of Cambridge (U.K.) is in physics, as is his 2006 Ph.D. from the Max Planck Institute in Munich (Germany), where he met and first began collaborating with Michael Wolf and David Pérez-García. Cubitt then spent the better part of a decade in mathematics departments, first at the University of Bristol (U.K.), then at Universidad Complutense de Madrid (Spain), before being awarded a Royal Society University Research Fellowship and returning to Cambridge. He completed the set by moving to the University College London department of computer science in 2015, where he is now a reader (associate professor) in quantum information and head of the CS Quantum group.

Cubitt is also a cofounder and director of the quantum start-up PhaseCraft. He was awarded the 2017 AHP-Birkhäuser Prize for research into connections between theoretical computer science and many-body quantum physics, which remains one of his major research interests.

Alessandro Di Bucchianico holds an M.Sc. in mathematics from Amsterdam University and a Ph.D. in mathematics from the University of Groningen, Groningen, Netherlands. Currently he is an associate professor of statistics in the Mathematics and Computing Science Department of Eindhoven University of Technology. From 2007 to 2011, he was the deputy head of LIME (Laboratory for Industrial Mathematics Eindhoven). He was quartermaster of the Data Science Center Eindhoven. From 2007 to 2019, he was the director of the permanent office of ENBIS, the European Network of Business and Industrial Statistics. From 1999 to 2009, he was one of the coordinators of the Industrial Statistics program at the research institute EURANDOM. At a national level, he is one of two coordinators of the Study Group Mathematics with Industry (http://www.swi-wiskunde.nl). He is a member of the Big Data team of the Applied Mathematics Institute of the 4 Technical Universities in the Netherlands—the activity that resulted in the current publication. His main research interests are statistical process control, reliability (both of hardware and software), and teaching of statistics.

Moon Duchin is a mathematician at Tufts University, where she's also a senior fellow in the Tisch College of Civic Life and the Director of the Program in Science, Technology, and Society. Since 2016, she has been focused on electoral redistricting as an application of mathematics to civil rights. In 2018 and 2019, she is working on interdisciplinary projects on gerrymandering as a Radcliffe Fellow and a Guggenheim Fellow.

Andrew Gelman is a professor of statistics and political science at Columbia University. He has published research articles on statistical theory, methods, and computation, with applications in social science and public health. He and his colleagues have written several books, including *Bayesian Data Analysis*, *Teaching Statistics: A Bag of Tricks*, *Regression and Other Stories*, *A Quantitative Tour of the Social Sciences*, and *Red State, Blue State, Rich State, Poor State: Why Americans Vote the Way They Do*.

Kevin Hartnett is the senior math writer for *Quanta Magazine*. From 2013 to 2016. he wrote "Brainiac," a weekly column for the *Boston Globe*'s Ideas section. A native of Maine, Kevin lives in South Carolina with his wife and three sons. Follow him on Twitter at @kshartnett.

Reuben Hersh is a magna cum laude graduate in English literature of Harvard College (1946), Ph.D. in math (NYU, 1962, Peter Lax), professor emeritus of the University of New Mexico, coauthor with Philip J Davis of *The Mathematical Experience* (National Book Award 1983), originator (with Richard J. Griego) of the theory and applications of random evolutions, author of *What Is Mathematics, Really?*, editor of *18 Unconventional Essays on the Nature of Mathematics*, author of *Experiencing Mathematics* (AMS) and *Peter Lax, Mathematician* (AMS), winner of the Chauvenet Prize (MAA) with Martin Davis and the Ford prize (MAA) with Ray Lorch. His article "Under-Represented Then Over-Represented on Jews in mathematics appeared in this series in 2011.

Theodore P. Hill is a professor emeritus of mathematics at Georgia Tech and lives on the Central Coast of California, where he is currently a Research Scholar in Residence at California Polytechnic State University. He is co-author of *An Introduction to Benford's Law* (Princeton University Press 2015), and author of the memoir *Pushing Limits: From West Point to Berkeley and Beyond* (AMS/MAA 2017). Most of Dr. Hill's 100-plus publications have been in mathematical probability, but he has also written essays for lay audiences that have appeared in *Academe, American Scientist* (five articles on different topics), *Chance, Chronicle of Higher Education, Mathematical Intelligencer*, and *Quillette*. For more information, see http://www.tphill.net.

Laura Iapichino holds a bachelor's degree in applied mathematics from the Università di Catania (Italy), an M.Sc. in applied mathematics from the Università degli Studi di Milano (Italy), and a Ph.D. in mathematics from the École polytechnique fédérale de Lausanne (EPFL) in Switzerland. Currently she is an assistant professor within the Scientific Computing Group at the Mathematics and Computing Science Department of Eindhoven University of Technology. She is a member of the Big Data team of the Applied Mathematics Institute of the 4 Technical Universities in the Netherlands—the activity that resulted in the current publication. Her research interests include model order reduction, parameterized partial differential equations, reduced basis methods, optimal control, optimization and multiobjective optimization, domain decomposition combined with reduced basis methods, computational fluid dynamics, and complex geometrical parameterization.

Vladimir Kanovei graduated in 1973 from Moscow State University, and obtained a Ph.D. in physics and mathematics from Moscow State University in 1976. In 1986, he became a doctor of science in physics and mathematics at

Moscow's Steklov Mathematical Institute (MIAN). He is currently the leading researcher at the Institute for Information Transmission Problems (IPPI), Moscow. Among his publications is the book *Borel Equivalence Relations: Structure and Classification*, University Lecture Series 44, American Mathematical Society, Providence, RI, 2008.

Karin U. Katz obtained her B.A. from Bryn Mawr College in 1982, and her Ph.D. from Indiana University in 1991. Her article "A Counterexample to the Equivariant Simple Loop Conjecture" (*Proc. Amer. Math. Soc.* 118 (1993), no. 1, 321–329) was based on her dissertation. One of her recent publications is the article "What Makes a Theory of Infinitesimals Useful? A View by Klein and Fraenkel" (with Vladimir Kanovei, Mikhail Katz, and Thomas Mormann) in *J. Humanist. Math.* 8 (2018), no. 1, 108–119.

Mikhail G. Katz (BA Harvard 1980; Ph.D. Columbia 1984) is a professor of mathematics at Bar Ilan University, Ramat Gan, Israel. His monograph *Systolic Geometry and Topology* was published by the American Mathematical Society. He is interested in Riemannian geometry, infinitesimals, debunking mathematical history written by the victors, as well as true infinitesimal differential geometry; see *Journal of Logic and Analysis* 7.5 (2015): 1–44 at http://dx.doi.org/10.4115/jla.2015.7.5. For a list of publications on the history, mathematics, and philosophy of inifnitesimals, see http://u.math.biu.ac.il/~katzmik/infinitesimals.html.

Semen S. Kutateladze (http://orcid.org/0000-0002-5306-2788) is a senior principal research officer of the Sobolev Institute of Mathematics in Novosibirsk, Russia, and professor at Novosibirsk State University. He has written more than 30 books and 300 papers on functional analysis, convex geometry, optimization, and nonstandard and Boolean valued analysis. He is a member of the editorial boards of *Siberian Mathematical Journal*, *Journal of Applied and Industrial Mathematics*, *Positivity*, and *Mathematical Notes*.

Mary Leng (https://orcid.org/0000-0001-9936-5453) is a senior lecturer in philosophy at the University of York (U.K.). Her book, *Mathematics and Reality* (Oxford U Press 2010), offers a defense of mathematical fictionalism, the view that we have no reason to believe that our mathematical theories are true. In 2016–2017, she held a research fellowship from the Leverhulme Trust to work on analogies between debates over realism and antirealism in the philosophy of mathematics and in metaethics. Perhaps surprisingly, she believes that we have more reason to be realists about moral truths than we have to be realists about the existence of mathematical objects.

Nelly Litvak is a professor at the University of Twente and Eindhoven University of Technology in the Netherlands. She received the Stieltjes Prize for the best Ph.D. thesis and the Google Faculty Research Award. Her research interests are in the study of large networks, such as online social networks and the World Wide Web, randomized algorithms, and random graphs. She is the leader of the Big Data team of the Applied Mathematics Institute of the 4 Technical Universities in the Netherlands—the activity that resulted in the current publication. In her free time, Nelly is a best-selling nonfiction author. Her book *Who Needs Mathematics? A Clear Book about How the Digital World Works*, coauthored with Andrei Raigorodski (in Russian, her native language), was short-listed for the 2017 Enlightener Prize for the best popular science book in Russia.

Melvyn B. Nathanson is a professor of mathematics at the City University of New York (Lehman College and the Graduate Center). An undergraduate major in philosophy at the University of Pennsylvania, he started graduate school at Harvard in biophysics, then began to study mathematics and completed a Ph.D. in mathematics at the University of Rochester in 1971. He was a visiting student at Cambridge University in 1970, worked with I. M. Gel'fand in Moscow under an IREX fellowship in 1972–1973, and was assistant to André Weil at the Institute for Advanced Study in 1974–1975. He is a Fellow of the American Mathematical Society.

Roice Nelson is a software developer with a passion for exploring mathematics through visualization. For him, to "see" something mathematical often means coding the ideas and rendering an image. A Platonist at heart, his mathematical interests center around non-Euclidean geometry, tilings, polytopes, and honeycombs. In 2017, he and his wife Sarah Nemec established an endowment at the University of Texas at Austin for mathematics outreach. He enjoys spending time with Sarah and their five cats; nurturing his Twitter bot, @TilingBot; unicycling; and indoor skydiving. Follow his mathematical adventures at roice3.org or on Twitter at @roice713.

Tahl Nowik is a professor of mathematics at Bar-Ilan University, Ramat Gan, Israel. His area of research is low-dimensional topology. He has studied finite order invariants of mappings in various low-dimensional settings, such as immersions of surfaces into 3-manifolds, curves in surfaces, and knot diagrams. He is interested in probabilistic questions in this setting and has studied random knots, random surfaces, and random simplicial complexes. He is also interested in nonstandard analysis, particularly in relation to differential geometry.

David Pérez-García started his research career working on functional analysis and obtained his Ph.D. in mathematics at Universidad Complutense de Madrid (Spain) in 2004. In 2005, he moved as a postdoctoral researcher to the Max Planck Institute of Quantum Optics in Germany, where he started to work in quantum information theory, his main research topic since then. Currently, he is a professor of mathematics at Universidad Complutense de Madrid and director of the research group Mathematics and Quantum Information. He has been awarded the Young Researcher RAC-Endesa Prize in Mathematics in 2012 and the Young Researcher Miguel Catalan Prize in 2017. In 2015, he was awarded an ERC Consolidator Grant to investigate the mathematics behind exotic quantum phases of matter.

James Propp is a full professor in the Department of Mathematical Sciences at the University of Massachusetts Lowell. Most of his research is in combinatorics, probability, and dynamical systems theory, with forays into the study of cellular automata and games. His monthly essays are posted at his Mathematical Enchantments blog at http://mathenchant.wordpress .com; one of these essays, "The Paintball Party," was published in *Math Horizons* and won the 2018 Trevor Evans Award from the Mathematical Association of America. Propp is a member of the advisory council of the National Museum of Mathematics and the chair of the advisory council of the Gathering 4 Gardner Foundation. You can follow him on Twitter at @jimpropp.

David M. Schaps (BA Swarthmore 1967, Ph.D. Harvard 1972) is a professor emeritus of classical studies at Bar-Ilan University, president of the Israel Society for the Promotion of Classical Studies (2016–2019), and author of *Economic Rights of Women in Ancient Greece*, *The Invention of Coinage and the Monetization of Ancient Greece*, *Handbook for Classical Research*, and a few dozen articles, including "The Woman Least Mentioned: Etiquette and Women's Names," "The Women of Greece in Wartime," "Zeus the Wife-Beater," and "Nausicaa the Comedienne: The *Odyssey* and the *Pirates of Penzance*."

David Sherry is a professor emeritus of philosophy at Northern Arizona University, among the tall pines of the Colorado Plateau. He has research interests in the philosophy of mathematics, especially applied mathematics. Recent publications include "Fields and the Intelligibility of Contact Action," *Philosophy* 90 (2015), 457–478; "Infinitesimals, Imaginaries, Ideals, and Fictions," with Mikhail Katz, *Studia Leibnitiana* 44 (2012), 166–192; and "Thermoscopes, Thermometers, and the Foundations of Measurement," *Studies in History and Philosophy of Science* 24 (2011), 509–524.

Neil J. A. Sloane (http://orcid.org/0000-0001-7386-5222) is the author of a dozen books, including *Sphere Packings, Lattices and Groups* (with J. H. Conway) and *The Theory of Error-Correcting Codes* (with F. J. MacWilliams), and more than 300 papers and videos. The (now On-Line) Encyclopedia of Integer Sequences (OEIS) was started by him in 1964 and contains more than 300,000 entries; he is the president of the OEIS Foundation. He is also a Member of the National Academy of Engineering, an AT&T Fellow, and a Fellow of the American Mathematical Society, the IEEE, and the Learned Society of Wales. Google Scholar Citations: 53,000, h-index: 79; i10-index: 249.

Kokichi Sugihara is a Meiji University distinguished professor emeritus, as well as a professor emeritus of the University of Tokyo. His research area is mathematical engineering. In his research on computer vision, he found a method for constructing 3D objects from "impossible figures" and extended his research interest to include human vision and optical illusion. Constructing mathematical models of human vision systems, he created various new impossible objects and won the first prize three times (2010, 2013, and 2018) and the second prize twice (2015 and 2016) in the Best Illusion of the Year Contest of the Neural Correlate Society.

Frank van der Meulen (https://orcid.org/0000-0001-7246-8612) is an associate professor in statistics at Delft University of Technology. His research is directed to statistical inference for stochastic processes, with a focus on uncertainty quantification and indirect observation schemes. Here, the stochastic process is defined as a continuous time object, but observations are discrete in time. Examples include Wiener driven stochastic differential equations and Lévy processes. He developed a general framework for conditional simulation of diffusion processes, which has been used in algorithms for Bayesian smoothing and parameter estimation. A second area of research consists of constructing algorithms for Bayesian nonparametric estimation and frequentist validation of Bayesian nonparametric inferential procedures. He is a member of the Big Data team of the Applied Mathematics Institute of the 4 Technical Universities in the Netherlands—the activity that resulted in the current publication.

Ron Wehrens is the business unit manager at Biometris, Wageningen University & Research, Wageningen, Netherlands. He earned his Ph.D. at Radboud University Nijmegen, Netherlands, and is the author of more than 100 research papers, one monograph, and several R packages available from CRAN and Bioconductor. His research interests include applied statistics, statistical software, chemometrics, and metabolomics.

Michael Wolf began his scientific career in high-energy physics. After a Ph.D. in mathematical physics under the supervision of Reinhard Werner and a postdoc at the Max Planck Institute of Quantum Optics, he became a professor at the Niels Bohr Institute in Copenhagen, Denmark, in 2008. In 2011, he moved to the Technical University of Munich, where he works as the chair of Mathematical Physics at the Department of Mathematics.

Noson S. Yanofsky has a Ph.D. in mathematics from The Graduate Center of The City University of New York. He is a professor of computer science at Brooklyn College and The Graduate Center. In addition to writing research papers, he has coauthored *Quantum Computing for Computer Scientists* (with Mirco A. Mannucci, Cambridge University Press) and authored *The Outer Limits of Reason: What Science, Mathematics, and Logic Cannot Tell Us* (MIT Press). He recently completed *Theoretical Computer Science for the Working Category Theorist* and is currently working on a book tentatively titled *Monoidal Categories: A Unifying Concept in Mathematics, Physics, and Computers*. Yanofsky lives in Brooklyn with his wife and four children.

Notable Writings

The following are other texts I considered for selection in this volume. As a space-saving rule, I did not include in this list articles published in the special journal issues which I also list—except for a very few pieces that deserve breaking the rule!

Akhlaghi-Ghaffarokh, Farbod. "Existence, Mathematical Nominalism, and Meta-Ontology: An Objection to Azzouni on Criteria for Existence." *Philosophia Mathematica* 26(2018): 251–65.

Akkach, Samer. "Aural Geometry: Poetry, Music, and Architecture in the Arabic Tradition." Pp. 166–95 in *Music, Sound, and Architecture in Islam*, edited by Michael Frishkofp and Frederico Spinetti. Austin, TX: University of Texas Press, 2018.

Alonso-Diaz, Santiago, and Jessica F. Cantlon. "Confidence Judgments during Ratio Comparisons Reveal a Bayesian Bias." *Cognition* 177(2018): 98–106.

Alshwaikh, Jehad. "Diagrams as Communication in Mathematics Discourse." *For the Learning of Mathematics* 38.2(2018): 9–13.

Ambrus, Gabriella, Andreas Filler, and Ödön Vancsó. "Functional Reasoning and Working with Functions: Functions/Mappings in Mathematics Teaching Tradition in Hungary and Germany." *Mathematics Enthusiast* 15.3(2018): 429–55.

Andersen, Holly. "Complements, Not Competitors: Causal and Mathematical Explanations." *British Journal for the Philosophy of Science* 69(2018): 485–508.

Asmuth, Jennifer, Emily M. Morson, and Lance J. Rips. "Children's Understanding of the Natural Numbers' Structure." *Cognitive Science* 42(2018): 1945–73.

Bailey, David. "Why Outliers Are Good for Science." *Significance* 15.1(2018): 14–18.

Bair, Jacques, Mikhail G. Katz, and David Sherry. "Fermat's Dilemma: Why Did He Keep Mum on Infinitesimals? And the European Theological Context." *Foundations of Science* 23(2018): 559–95.

Baldwin, John T. "Axiomatizing Changing Conceptions of the Geometric Continuum I: Euclid-Hilbert." *Philosophia Mathematica* 26(2018): 346–74.

Barany, Michael J. "Integration by Parts: Wordplay, Abuses of Language, and Modern Mathematical Theory on the Move." *Historical Studies in the Natural Sciences* 48.3(2018): 259–99.

Barba, Kimberly. "The Portrayal of Mathematicians and Mathematics in Popular Culture." *Journal of Mathematics Education at Teachers College* 9.1(2018): 9–14.

Barnett, Arnold. "Epic Fail? The Polls and the 2016 Presidential Election." *Chance* 31.4(2018): 4–8.

Barton, Craig. "On Formative Assessment in Math." *American Educator* 42.2(2018): 33–43.

Bascelli, Tiziana, et al. "Cauchy's Infinitesimals, His Sum Theorem, and Foundational Paradigms." *Foundations of Science* 23(2018): 267–96.

Bathfield, Maël. "Why Zeno's Paradoxes of Motion are Actually about Immobility." *Foundations of Science* 23(2018): 649–79.

Benci, Vieri, Leon Horsten, and Sylvia Wenmackers. "Infinitesimal Probabilities." *British Journal for the Philosophy of Science* 69(2018): 509–52.

Berry, Don. "Proof and the Virtues of Shared Enquiry." *Philosophia Mathematica* 26(2018): 112–30.

Biggs, Norman. "Game, Set, and Graph." *BSHM Bulletin—Journal of the British Society for the History of Mathematics* 33.3(2018): 166–78.

Bitbol, Michel. "Mathematical Demonstration and Experimental Activity: A Wittgensteinian Philosophy of Physics." *Philosophical Investigations* 41.2(2018): 188–203.

Blåsjö, Viktor. "Mathematicians versus Philosophers in Recent Work on Mathematical Beauty." *Journal of Humanistic Mathematics* 8.1(2018): 414–31.

Boggs, George, et al. "Contextual Meanings of the Equals Sign as Conceptual Blends." *For the Learning of Mathematics* 38.2(2018): 34–39.

Bolton, Alexander. "Topological Tic-Tac-Toe." *Chalkdust Magazine* 8(Autumn 2018): 49–53.

Bonnai, Denis, and Jacques Dubucs. "Philosophy of Mathematics." Pp. 349–404 in *The Philosophy of Science: A Companion*, edited by Anouk Barberousse, Denis Bonai, and Mikaël Cozic. New York: Oxford University Press, 2018.

Bonner, Jay. "Doing the Jitterbug with Islamic Geometric Patterns." *Journal of Mathematics and the Arts* 12.2/3(2018): 128–43.

Bornn, Luke, Dan Cervone, and Javier Fernandez. "Soccer Analytics: Unravelling the Complexity of 'the Beautiful Game.'" *Significance* 15.3(2018): 26–29.

Bracken, A. J. "Mathematics Underfoot: The Formulas That Came to Würzburg from New Haven." *The Mathematical Intelligencer* 40.2(2018): 67–75.

Breen, Richard, Kristian Bernt Karlson, and Anders Holm. "Interpreting and Understanding Logits, Probits, and Other Nonlinear Probability Models." *Annual Review of Sociology* 44(2018): 39–54.

Bressan, Paola. "Systemisers *Are* Better at Math." *Nature Scientific Reports* Aug. 2, 2018.

Briggs, William. "Quantitative Literacy and Civic Virtue." *Numeracy* 11.2(2018).

Brooks, Neon B., et al. "The Role of Gesture in Supporting Mental Representations: The Case of Mental Abacus Arithmetic." *Cognitive Science* 42(2018): 554–75.

Butterworth, Brian. "Mathematical Expertise." Pp. 616–33 in *Cambridge Handbook of Expertise and Expert Performance*, 2nd ed., edited by Anders K. Ericsson, Robert T. Hoffman, Aaron Kozbelt, and Mark A. Williams. New York: Cambridge University Press, 2018.

Byers, William. "Can You Say What Mathematics Is?" Pp. 45–57 in *Humanizing Mathematics and Its Philosophy*, edited by Bharat Sriraman. Cham, Switzerland: Springer International Publishing, 2017.

Campos, Daniel G. "Heuristic Analogy in *Ars Conjectandi*: From Archimedes' *De Circuli Dimensione* to Bernoulli's Theorem." *Studies in History and Philosophy of Science, A* 67(2018): 44–53.

Caraman, Sânziana, and Lorelei Caraman. "Brouwer's Fixed Point Theorem and the Madeleine Moment." *Journal of Mathematics and the Arts* 12.4(2018): 207–24.

Carvajalino, Juan. "Edwin Bidwell Wilson and Mathematics as a Language." *Isis* 109.3(2018): 494–514.

Chanier, Thomas. "A Possible Solution to the Mayan Calendar Enigma." *The Mathematical Intelligencer* 40.3(2018): 18–25.

Chen, Jiang-Ping Jeff. "A Systematic Treatment of 'Linear Algebra' in 17th-Century China." *The College Mathematics Journal* 49.3(2018): 169–79.

Chu, Pingyi. "A Mirror to the Calendar: Cut, Paste and Spread the European Calendrical Learning." Pp. 351–66 in *Visual and Textual Representations in Exchanges between Europe and East Asia, 16th–18th Centuries*, edited by Luís Saraiva and Catherine Jami. Singapore: World Scientific, 2018.

Clader, Emily. "Why Twelve Tones? The Mathematics of Musical Tuning." *The Mathematical Intelligencer* 40.3(2018): 32–36.

Cobb, Sarah C., and Jeff B. Hood. "Mathematical Arguments in Favor of Risk in Andy Weir's *The Martian*." *Journal of Humanistic Mathematics* 8(2018).

Cokely, Edward T., et al. "Skilled Decision Theory: From Intelligence to Numeracy and Expertise." Pp. 476–505 in *Cambridge Handbook of Expertise and Expert Performance*, 2nd ed., edited by Anders K. Ericsson, Robert T. Hoffman, Aaron Kozbelt, and Mark A. Williams. New York: Cambridge University Press, 2018.

Colyvan, Mark. "The Ins and Outs of Mathematical Explanation." *The Mathematical Intelligencer* 40.4(2018): 26–29.

Colyvan, Mark, John Cusbert, and Kelvin McQueen. "Two Flavours of Mathematical Explanation." Pp. 231–49 in *Explanation beyond Causation: Philosophical Perspectives on Non-Causal Explanations*, edited by Alexander Reutlinger and Juha Saati. Oxford, U.K.: Oxford University Press, 2018.

Combs, Randy, Teri Bingham, and Taylor Roper. "A Model for Inverting the Advanced Calculus Classroom." *PRIMUS* 28.8(2018): 717–25.

Confalonieri, Sara. "A Further Analysis of Cardano's Main Tool in the *De Regula Aliza*: On the Origins of the Splittings." *Archive for History of Exact Sciences* 72(2018): 303–52.

Costello, Fintan, and Paul Watts. "Invariants in Probabilistic Reasoning." *Cognitive Psychology* 100(2018): 1–15.

Costello, Fintan, Paul Watts, and Christopher Fisher. "Surprising Rationality in Probability Judgment: Assessing Two Competing Models." *Cognition* 170(2018): 280–97.

Craig, Jeffrey. "The Promises of Numeracy." *Educational Studies in Mathematics* 99(2018): 57–71.

Craig, Jeffrey, and Lynette Guzmán. "Six Propositions of a Social Theory of Numeracy." *Numeracy* 11.2(2018).

Dakić, Branimir, and Biserka Kolarec. "A Mathematician in Zagreb." *The Mathematical Intelligencer* 40.4(2018): 19–22.

D'Alessandro, William. "Arithmetic, Set Theory, Reduction and Explanation." *Synthese* 195(2018): 5059–89.

Darragh, Lisa. "Loving and Loathing: Portrayals of School Mathematics in Young Adult Fiction." *Journal for Research in Mathematics Education* 49.2(2018): 178–209.

Date-Huxtable, Elizabeth, et al. "Conceptualisations of Infinity by Primary Pre-Service Teachers." *Mathematics Education Research Journal* 30(2018): 545–67.

Davis, Brent. "What Sort of Science Is Didactics?" *For the Learning of Mathematics* 38.3(2018): 44–49.

de Freitas, Elizabeth. "The Mathematical Continuum: A Haunting Problem." *The Mathematical Enthusiast* 15.1–2(2018): 148–58.

de Ronde, Christian. "Quantum Superpositions and the Representation of Physical Reality beyond Measurement Outcomes and Mathematical Structures." *Foundations of Science* 23(2018): 621–48.

de Rouilhan, Philippe. "Philosophy of Logic." Pp. 319–48 in *The Philosophy of Science: A Companion*, edited by Anouk Barberousse, Denis Bonai, and Mikaël Cozic. New York: Oxford University Press, 2018.

DeDieu, Lauren, and Miroslav Lovric. "Student Perceptions of the Use of Writing in a Differential Equations Course." *PRIMUS* 28.2(2018): 166–85.

Del Centina, Andrea, and Alessandra Fiocca. "'A Masterly though Neglected Work,' Boscovich's Treatise on Conic Sections." *Archive for History of Exact Sciences* 72(2018): 453–95.

Del Centina, Andrea, and Alessandra Fiocca. "Boscovich's Geometrical Principle of Continuity and the 'Mysteries of Infinity.'" *Historia Mathematica* 45(2018): 131–75.

Drekalović, Vladimir, and Berislav Žarnić. "Which Mathematical Objects Are Referred to by the Enhanced Indispensability Argument?" *Journal for General Philosophy of Science* 49(2018): 121–26.

Dreyfus, Tommy. "Learning through Activity: Basic Research on Mathematical Cognition." *The Journal of Mathematical Behavior* 52(2018): 216–23.

Duklewski, Ben, Veera Holdai, and Barbara Wainwright. "Some Limitations of Financial Models." *Undergraduate Mathematics and Its Applications* 39.1(2018): 41–68.

Dunham, William. "The Early (and Peculiar) History of the Möbius Function." *Mathematics Magazine* 91.2(2018): 83–91.

Dyson, George. "Childhood's End." *Edge Online* Jan. 1, 2019.

Elior, Ofer. "The Arabic Tradition of Euclid's Elements Preserved in the Latin Translation by Adelard de Bath and in the Hebrew Translation by Rabbi Jacob." *Historia Mathematica* 45(2018): 111–30.

Ellis, George F. R., Krzysztof A. Meissner, and Hermann Nicolai. "The Physics of Infinity." *Nature Physics* 14(2018): 770–72.

Estrada, Ernesto. "Integer-Digit Functions: An Example of Math-Art Integration." *The Mathematical Intelligencer* 40.1(2018): 73–78.

Everett, Jonathan. "A Kantian Account of Mathematical Modelling and the Rationality of Scientific Theory Change." *Studies in History and Philosophy of Science, A* 71(2018): 45–57.

Feintzeig, Benjamin H. "On the Choice of Algebra for Quantization." *Philosophy of Science* 85.1(2018): 102–25.

Frans, Joachim, Isar Goyvaerts, and Bart van Kerkhove. "Model-Based Reasoning in Mathematical Practice." Pp. 537–49 in *Springer Handbook of Model-Based Science*, edited by Lorenzo Magnani and Tommaso Bertolotti. Dordrecht, Germany: Springer International, 2017.

Frantz, Marc. "The ABCs of Viewing Ringed Planets." *The Mathematical Intelligencer* 40.1 (2018): 4–13.

Fried, Michael N. "Ways of Relating to the Mathematics of the Past." *Journal of Humanistic Mathematics* 8.1(2018).

Futamura, Fumiko, and Alison M. Marr. "Taking Mathematics Abroad: A How-To Guide." *PRIMUS* 28.9(2018): 875–89.

Gal, Kobi, et al. "Which Is the Fairest (Rent Division) of Them All?" *Communications of the ACM* 61.1(2018): 93–100.

Geary, David C., and Kristy van Marle. "Growth of Symbolic Number Knowledge Accelerates after Children Understand Cardinality." *Cognition* 177(2018): 69–78.

Ghomi, Muhammad. "Dürer's Unfolding Problem for Convex Polyhedra." *Notices of the American Mathematical Society* 65.1(2018): 25–27.

Giardino, Valeria. "Diagrammatic Reasoning in Mathematics." Pp. 499–522 in *Springer Handbook of Model-Based Science*, edited by Lorenzo Magnani and Tommaso Bertolotti. Dordrecht, Germany: Springer International, 2017.

Gill, Richard D., Piet Groeneboom, and Peter de Jong. "Elementary Statistics on Trial: The Case of Lucia de Berk." *Chance* 31.4(2018): 9–15.

Ginovart, Josep Lluis i, Mónica López-Piquer, and Judith Urbano-Lorente. "Transfer of Mathematical Knowledge for Building Medieval Cathedrals." *Nexus Network Journal* 20(2018): 153–72.

Goodwin, William. "Conflicting Conceptions of Construction in Kant's Philosophy of Geometry." *Perspectives on Science* 26.1(2018): 97–118.

Goren, Nurullah E., and Tiffany Zhu. "The Humanistic Mathematics Network Journal: A Bibliographic Report." *Journal of Humanistic Mathematics* 8.1(2018).

Gorroochurn, Prakash. "The End of Statistical Independence: The Story of Bose–Einstein Statistics." *The Mathematical Intelligencer* 40.3(2018): 12–17.

Grant, Robert. "Calculate *and* Communicate." *Significance* 15.6(2018): 42–44.

Griffths, Thomas L., et al. "Subjective Randomness as Statistical Inference." *Cognitive Psychology* 103(2018): 85–109.

Guerrero-Ortiz, Carolina, Jaime Mena-Lorca, and Astrid Morales Soto. "Fostering Transit between Real World and Mathematical World: Some Phases on the Modelling Cycle." *International Journal of Science and Mathematics Education* 16(2018): 1605–28.

Guicciardini, Niccolò. "Un Altro Presente: On the Historical Interpretation of Mathematical Texts." *BSHM Bulletin: Journal of the British Society for the History of Mathematics* 33.3(2018): 148–65.

Hansen, Casper Storm. "Two Envelopes and Binding." *Australasian Journal of Philosophy* 96.3 (2018): 508–18.

Heule, Marijn J. H. "Brute Trust." *Nieuw Archief voor Wiskunde* 19.3(2018): 226–27.

Hicks, Stephanie C., and Rafael A. Irizarry. "A Guide to Teaching Data Science." *The American Statistician* 72.4(2018): 382–91.

Hisarligil, Hakan, and Beyhan Bolak Hisarligil. "The Geometry of Cuboctahedra in Medieval Art in Anatolia." *Nexus Network Journal* 20(2018): 125–52.

Holik, Federico, et al. "Pattern Recognition in Non-Kolmogorovian Structures." *Foundations of Science* 23(2018): 119–32.

Howson, Colin. "The Curious Case of Frank Ramsey's Proof of the Multiplication Rule of Probability." *Analysis* 78.3(2018): 431–39.

Icard, Thomas F. "Bayes, Bounds, and Rational Analysis." *Philosophy of Science* 85.1(2018): 79–101.

Inglis, Matthew, and Colin Foster. "Five Decades of Mathematics Education Research." *Journal for Research in Mathematics Education* 49.4(2018): 462–500.

Isaac, Manuel Gustavo. "Toward a Phenomenological Epistemology of Mathematical Logic." *Synthese* 195(2018): 863–74.

Jardine, Boris. "Instruments of Statecraft: Humphrey Cole, Elizabethan Economic Policy and the Rise of Practical Mathematics." *Annals of Science* 75.4(2018): 304–29.

Jenny, Matthias. "Counterpossibles in Science: The Case of Relative Computability." *Noûs* 52.3(2018): 530–60.

Jones, Steven R. "Prototype Images in Mathematics Education: The Case of the Graphical Representation of the Definite Integral." *Educational Studies in Mathematics* 97(2018): 215–34.

Kanjlia, Shipra, Lisa Feigenson, and Marina Bedny. "Numerical Cognition Is Resilient to Dramatic Changes in Early Sensory Experience." *Cognition* 179(2018): 111–20.

Karam, Ricardo. "Fresnel's Original Interpretation of Complex Numbers in 19th Century Optics." *American Journal of Physics* 86.4(2018): 245–49.

Katz, Boris, Mikhail G. Katz, and Sam Sanders. "A Footnote to 'The Crisis in Contemporary Mathematics.'" *Historia Mathematica* 45(2018): 176–81.

Kawan, Jamhari, and Waranat Wongkia. "Experiencing the Angle Properties in a Circle." *Australian Mathematics Teacher* 74.3(2018): 24–33.

Kenan, Kok Xiao-Feng. "Igniting the Joy of Learning Mathematics." *Australian Mathematics Teacher* 74.3(2018): 34–40.

Kim, Dan, and John E. Opfer. "Dynamics and Development in Number-to-Space Mapping." *Cognitive Psychology* 107(2018): 44–66.

Knappik, Franz. "Bayes and the First Person: Consciousness of Thoughts, Inner Speech and Probabilistic Inference." *Synthese* 195(2018): 2113–40.

Kollosche, David. "Social Functions of Mathematics Education: A Framework for Socio-Political Studies." *Educational Studies in Mathematics* 98(2018): 287–303.

Kollosche, David. "The True Purpose of Mathematics Education: A Provocation." *Mathematics Enthusiast* 15.1/2(2018): 303–20.

Komarova, Mariya. "World Bobsleigh Tracks: From Geometry to the Architecture of Sports Facilities." *Nexus Network Journal* 20.1(2018): 235–49.

Konoval, Brandon. "Pythagorean Pipe Dreams? Vincenzo Galilei, Marin Mersenne, and the Pneumatic Mysteries of the Pipe Organ." *Perspectives on Science* 26.1(2018): 1–51.

Kontorovich, Igor'. "Undergraduates' Images of the Root Concept in $\mathbb{R}$ and in $\mathbb{C}$." *The Journal of Mathematical Behavior* 49(2018): 184–93.

Krupnik, Victoria, Timothy Fukawa-Connelly, and Keith Weber. "Students' Epistemological Frames and Their Interpretation of Lectures in Advanced Mathematics." *The Journal of Mathematical Behavior* 49(2018): 174–83.

Lampert, Timm. "Wittgenstein and Gödel: An Attempt to Make 'Wittgenstein's Objection' Reasonable." *Philosophia Mathematica* 26(2018): 324–45.

Lange, Marc. "Mathematical Explanations That Are Not Proofs." *Erkenntnis* 83(2018): 1285–302.

Langkjær-Bain, Robert. "Five Ways Data Is Transforming Music." *Significance* 15.1(2018): 20–23.

Langkjær-Bain, Robert. "Where the Seeds of Modern Statistics Were Sown." *Significance* 15.3(2018): 14–19.

Lao, Limin, and Jennifer Hall. "The Important Things about Writing in Secondary Mathematics Classes." *Australian Mathematics Teacher* 74.1(2018): 13–19.

Lapointe, Sandra. "Bolzano on Logic in Mathematics and Beyond." Pp. 101–22 in *Logic from Kant to Russell: Laying the Foundations for Analytic Philosophy*, edited by Sandra Lapointe. New York: Routledge, 2019.

Larsen, Sean. "Didactical Phenomenology: The Engine That Drives Realistic Mathematics Education." *For the Learning of Mathematics* 38.3(2018): 25–29.

Lazer, David, and Jason Radford. "Data ex Machina: Introduction to Big Data." *Annual Review of Sociology* 43(2017): 19–39.

Lê, François. "The Recognition and the Constitution of the Theorems of Closure." *Historia Mathematica* 45(2018): 237–76.

Lenhard, Johannes, and Michael Otte. "The Applicability of Mathematics as a Philosophical Problem: Mathematization as Exploration." *Foundations of Science* 23(2018): 719–37.

Lerman, Stephen. "Towards Subjective Truths in Mathematics Education." *For the Learning of Mathematics* 38.3(2018): 54–56.

Linnebo, Øystein. "Truth in Mathematics." Pp. 648–65 in *The Oxford Handbook of Truth*, edited by Michael Glanzberg. Oxford, U.K.: Oxford University Press, 2018.

Louie, Nicole L. "Culture and Ideology in Mathematics Teacher Noticing." *Educational Studies in Mathematics* 97(2018): 55–69.

Lucente, Sandra, and Antonio Macchia. "A Zen Master, a Zen Monk, a Zen Mathematician." *Nexus Network Journal* 20(2018): 459–74.

Luecking, Stephen. "A Toroidal Walk in the Park." *Math Horizons* 25.1(2017): 12–14.

Lutovac, Sonja, and Raimo Kaasila. "Future Directions in Research on Mathematics-Related Teacher Identity." *International Journal of Science and Mathematics Education* 16(2018): 759–76.

Lyons, Emily McLaughlin, et al. "Stereotype Threat Effects on Learning from a Cognitively Demanding Mathematics Lesson." *Cognitive Science* 42(2018): 678–90.

MacKinnon, Edward. "The Role of a Posteriori Mathematics in Physics." *Studies in History and Philosophy of Modern Physics* 62(2018): 166–75.

Mahajan, Sanjoy. "A Quantity without Its Units Is Like Water without Wetness." *American Journal of Physics* 86.5(2018): 381–83.

Mahajan, Sanjoy. "The Exponential Benefits of Logarithmic Thinking." *American Journal of Physics* 86.11(2018): 859–61.

Maheux, Jean-François, and Jérôme Proulx. "Introduction." *Mathematics Enthusiast* 15.1/2 (2018): 78–99.

Mahtani, Anna. "Imprecise Probabilities and Unstable Betting Behavior." *Noûs* 52.1(2018): 69–87.

Mamary, Anne. "Geometry in the Humming of the Strings." Pp. 3–18 in *The Bright and the Good: The Connection between Intellectual and Moral Virtues*, edited by Audrey L. Anton. Lanham, MD: Rowman & Littlefield, 2018.

Mancosu, Paolo. "The Origin of the Group in Logic and the Methodology of Science." *Journal of Humanistic Mathematics* 8.1(2018).

Marshall, Oliver R. "The Psychology and Philosophy of Natural Numbers." *Philosophia Mathematica* 26(2018): 40–58.

Martín-Molina, Verónica, et al. "Researching How Professional Mathematicians Construct New Mathematical Definitions: A Case Study." *International Journal of Mathematical Education in Science and Technology* 49.7(2018): 1069–82.

Matherne, Samantha. "Merleau-Ponty on Abstract Thought in Mathematics and Natural Science." *European Journal of Philosophy* 26(2018): 780–97.

McKenna, Douglas M. "On a Better Golden Rectangle (That Is Not 61.8033 . . . Useless!)" Pp. 187–94 in *Proceedings of Bridges 2018: Mathematics, Art, Music, Architecture, Education, Culture*, edited by Eve Torrence, Bruce Torrence, Carlo H. Séquin, and Kristóf Fenyvesi. Phoenix: Tessellations Publishing, 2018.

Menken, Jane. "On Becoming a Mathematical Demographer." *Annual Review of Sociology* 44 (2018): 1–17.

Mohanty, J. N. "Philosophy of Logic." *Journal of Indian Council of Philosophical Research* 35 (2018): 3–14.

Moulton, Derek E., Alain Goriely, and Régis Chirat. "How Seashells Take Shape." *Scientific American* 318.4(2018): 69–75.

Mueller, Julia. "On the Genesis of Robert P. Langlands' Conjectures and His Letter to André Weil." *Bulletin of the American Mathematical Society* 55.4(2018): 493–528.

Mumma, John. "Deduction, Diagrams, and Model-Based Reasoning." Pp. 523–35 in *Springer Handbook of Model-Based Science*, edited by Lorenzo Magnani and Tommaso Bertolotti. Dordrecht, Germany: Springer International, 2017.

Mynard, Frédéric. "Distinguishing the Plane from the Punctured Plane without Homotopy." *Mathematics Magazine* 91.1(2018): 16–19.

Nabb, Keith, et al. "Using the 5 Practices in Mathematics Teaching." *Mathematics Teacher* 111.5(2018): 366–73.

Ng, Chi-hung Clarence. "High School Students' Motivation to Learn Mathematics: The Role of Multiple Goals." *International Journal of Science and Mathematics Education* 16(2018): 357–75.

Ng, Oi-Lam, Nathalie Sinclair, and Brent Davis. "Drawing Off the Page: How New 3D Technologies Provide Insight into Cognitive and Pedagogical Assumptions about Mathematics." *The Mathematical Enthusiast* 15.3(2018): 563–77.

Niss, Martin. "A Mathematician Doing Physics: Mark Kac's Work on the Modeling of Phase Transitions." *Perspectives on Science* 26.2(2018): 185–212.

Nordgren, Ronald P. "How Franklin (May Have) Made His Squares." *Mathematics Magazine* 91.1(2018): 24–32.

Norton, Anderson. "Frameworks for Modeling Students' Mathematics." *The Journal of Mathematical Behavior* 52(2018): 201–7.

Oaks, Jeffrey A. "François Viète's Revolution in Algebra." *Archive for History of Exact Sciences* 72(2018): 245–302.

Ohtani, Hiroshi. "Philosophical Pictures about Mathematics: Wittgenstein and Contradiction." *Synthese* 195(2018): 2039–63.

O'Malley, Maureen A., and Emily C. Parke. "Microbes, Mathematics, and Models." *Studies in History and Philosophy of Science, A* 72(2018): 1–10.

Ornes, Stephen. "Art by Numbers." *Scientific American* 319.2(2018): 69–73.

Ornes, Stephen. "It's All Smooths and Garns." *Physics World Online* Nov. 20, 2018.

Ornes, Stephen. "Math Tools Send Legislators Back to the Drawing Board." *Proceedings of the National Academy of Sciences* 115.26(2018): 6515–17.

Ossendrijver, Mathieu. "Bisecting the Trapezoid: Tracing the Origins of a Babylonian Computation of Jupiter's Motion." *Archive for History of Exact Sciences* 72(2018): 145–89.

Otàrola-Castillo, Erik, and Melissa G. Torquato. "Bayesian Statistics in Archaeology." *Annual Review of Anthropology* 47(2018): 435–53.

Pak, Igor. "How to Write a Clear Math Paper: Some 21st Century Tips." *Journal of Humanistic Mathematics* 8.1(2018).

Papageorgiou, Eleni, and Constantinos Xenofontos. "Discovering Geometrical Transformations in the Ancient Mosaics of Cyprus." *Australian Mathematics Teacher* 74.2(2018): 34–40.

Parberry, Ian. "The Unexpected Beauty of Modular Bivariate Quadratic Functions." *Journal of Mathematics and the Arts* 12.4(2018): 197–206.

Patel, Purav, and Sashank Varma. "How the Abstract Becomes Concrete: Irrational Numbers Are Understood Relative to Natural Numbers and Perfect Squares." *Cognitive Science* 42(2018): 1642–76.

Peressini, Anthony F. "Causation, Probability, and the Continuity Bind." *British Journal for the Philosophy of Science* 69(2018): 881–909.

Pettersen, Andreas, and Guri A. Nortvedt. "Identifying Competency Demands in Mathematical Tasks: Recognising What Matters." *International Journal of Science and Mathematics Education* 16(2018): 949–65.

Piazza, Manuela, Vito De Feo, Stefano Panzeri, and Stanislas Dehaene. "Learning to Focus on Number." *Cognition* 181(2018): 35–45.

Pieronkiewicz, Barbara. "Mathematicians Who Never Were." *The Mathematical Intelligencer* 40.2(2018): 45–49.

Planas, Núria. "Language as Resource: A Key Notion for Understanding the Complexity of Mathematics Learning." *Educational Studies in Mathematics* 98(2018): 915–29.

Polster, Burkard, and Marty Ross. "Marching in Squares." *The College Mathematics Journal* 49.3(2018): 181–86.

Porter, Theodore M. "Politics by the Numbers." Pp. 23–34 in *Education by the Numbers and the Making of Society: The Expertise of International Assessments*, edited by Sverker Lindblad, Daniel Petterson, and Thomas S. Popkewitz. Abingdon, U.K.: Routledge, 2018.

Pringe, Hernán. "Maimon's Criticism of Kant's Doctrine of Mathematical Cognition and the Possibility of Metaphysics as a Science." *Studies in History and Philosophy of Science, A* 71(2018): 35–44.

Propp, James. "The Paintball Party." *Math Horizons* 25.2(2017): 18–21.

Quintana, Federico Raffo. "Leibniz on the Requisites of an Exact Arithmetical Quadrature." *Studies in History and Philosophy of Science, A* 67(2018): 65–73.

Rabouin, David. "Logic of Imagination: Echoes of Cartesian Epistemology in Contemporary Philosophy of Mathematics and Beyond." *Synthese* 195(2018): 4751–83.

Radovic, Darinka, et al. "Towards Conceptual Coherence in the Research on Mathematics Learner Identity." *Educational Studies in Mathematics* 99(2018): 21–42.

Rahaman, Jeenath, et al. "Recombinant Enaction: Manipulatives Generate New Procedures in the Imagination, by Extending and Recombining Action Spaces." *Cognitive Science* 42(2018): 370–415.

Rahmatian, Sasan. "The Structure of Argentine Tango." *Journal of Mathematics and the Arts* 12.4(2018): 225–43.

Raper, Simon. "Bernoulli's Golden Theorem." *Significance* 15.4(2018): 26–29.

Reeder, Patrick. "Labyrinth of Continua." *Philosophia Mathematica* 26(2018): 1–39.

Reese, Allan R. "Graphical Interpretations of Data." *Significance,* six different issues, 2017–18.

Reimers, Stian, Chris Donkin, and Mike E. Le Pelley. "Perceptions of Randomness in Binary Sequences: Normative, Heuristic, or Both." *Cognition* 172(2018): 11–25.

Reinholz, Daniel L. "Large Lecture Halls: Whiteboards, Not Bored Students." *PRIMUS* 28.7(2018): 670–82.

Reinholz, Daniel L. "Peer Feedback for Learning Mathematics." *The American Mathematical Monthly* 125.7(2018): 653–58.

Repplinger, Michael, Lisa Beinborn, and Willem Zuidema. "Vector-Space Models of Words and Sentences." *Nieuw Archief voor Wiskunde* 5/19.3(2018): 167–74.

Rian, Iasef Md. "Fractal-Based Computational Modeling and Shape Transition of a Hyperbolic Paraboloid Shell Structure." *Nexus Network Journal* 20(2018): 437–58.

Rieger, Adam. "The Beautiful Art of Mathematics." *Philosophia Mathematica* (2018): 234–50.

Rivera, Ferdinand. "Abduction and the Emergence of Necessary Mathematical Knowledge." Pp. 551–67 in *Springer Handbook of Model-Based Science*, edited by Lorenzo Magnani and Tommaso Bertolotti. Dordrecht, Germany: Springer International, 2017.

Rizza, Davide. "A Study of Mathematical Determination through Bertrand's Paradox." *Philosophia Mathematica* 26(2018): 375–95.

Rodal, Jocelyn. "Patterned Ambiguities: Virginia Woolf, Mathematical Variables, and Form." *Configurations* 26.1(2018): 73–101.

Rolla, Krishna Priya. "Human Capital: The Mathematics of Measurement." Pp. 345–84 in *Human Capital and Assets in the Networked World*, edited by Meir Russ. Bingley, U.K.: Emerald Publishing, 2017.

Romero-Vallhonesta, Fàtima, and M. Rosa Massa-Esteve. "The Main Sources for the *Arte Mayor* in Sixteenth Century Spain." *BSHM Bulletin—Journal of the British Society for the History of Mathematics* 33.2(2018): 73–95.

Rowe, David E. "On Models and Visualizations of Some Special Quartic Surfaces." *The Mathematical Intelligencer* 40.1(2018): 59–67.

Ruge, Johanna. "On Epistemological Violence in Mathematics Education Research." *Mathematics Enthusiast* 15.1/2(2018): 320–44.

Ruiz-Primo, Maria Araceli, and Heidi Kroog. "Looking Closely at Mathematics and Science Classroom Feedback Practices." Pp. 191–218 in *The Cambridge Handbook of Instructional Feedback*, edited by Anastasiya A. Lipnevich and Jeffrey K. Smith. Cambridge, U.K.: Cambridge University Press, 2019.

Runyan, Jason D. "Agent-Causal Libertarianism, Statistical Neural Laws and Wild Coincidences." *Synthese* 195(2018): 4563–80.

Ryazanov, Arseny A., et al. "Intuitive Probabilities and the Limitation of Moral Imagination." *Cognitive Science* 42 suppl(2018): 38–68.

Saatsi, Juha. "On Explanations from Geometry of Motion." *British Journal for the Philosophy of Science* 69(2018): 253–73.

Sakkal, Mamoun. "Intersecting Squares: Applied Geometry in the Architecture of Timurid Samarkand." *Journal of Mathematics and the Arts* 12.2/3(2018): 65–95.

Sandoz, Raphaël. "Applying Mathematics to Empirical Sciences: Flashback to a Puzzling Disciplinary Interaction." *Synthese* 195(2018): 875–98.

Santolin, Chiara, and Jenny R. Saffran. "Constraints on Statistical Learning across Species." *Trends in Cognitive Sciences* 22.1(2018): 522–63.

Savage, Neil. "Always Out of Balance." *Communications of the ACM* 61.4(2018): 12–14.

Savage, Neil. "Using Functions for Easier Programming." *Communications of the ACM* 61.5 (2018): 29–30.

Sawatzki, Carly, and Peter Sullivan. "Shopping for Shoes: Teaching Students to Apply and Interpret Mathematics in the Real World." *International Journal of Science and Mathematics Education* 16(2018): 1355–73.

Sawyer, Kim. "Synchronicity and the Search for Significance." *Significance* 15.6(2018): 34–37.

Schattschneider, Doris. "Marjorie Rice and the MAA Tiling." *Journal of Mathematics and the Arts* 12.2/3(2018): 114–27.

Schneider, Alyse. "Gramsci's Contradictions in Mathematics Education Researcher Positionality." *Mathematics Enthusiast* 15.1/2(2018): 100–33.

Schwarz, Wolfgang. "No Interpretation of Probability." *Erkenntnis* 83(2018): 1195–212.

Séquin, Carlo H. "Möbius Bridges." *Journal of Mathematics and the Arts* 12.2/3(2018): 181–94.

Sevimli, Eyup. "Understanding Students' Hierarchical Thinking: A View from Continuity, Differentiability and Integrability." *Teaching Mathematics and its Applications* 37.1(2018): 1–16.

Shaki, Samuel, and Martin H. Fischer. "Deconstructing Spatial-Numerical Associations." *Cognition* 175(2018): 109–13.

Shelburne, Brian J. "The Goldilocks of Number Systems." *Math Horizons* 25.4(2018): 10–13.

Shen, Alexander. "Hilbert's Error?" *The Mathematical Intelligencer* 40.4(2018): 6–11.

Sherry, David. "The Jesuits and the Method of Indivisibles." *Foundations of Science* 23(2018): 367–92 [followed by an Amir Alexander comment].

Sidoli, Nathan. "The Concept of *Given* in Greek Mathematics." *Archive for History of Exact Sciences* 72(2018): 353–402.

Sidoli, Nathan. "Uses of Construction in Problems and Theorems in Euclid's *Elements* I–VI." *Archive for History of Exact Sciences* 72(2018): 403–52.

Sidoli, Nathan. "Mathematics Education." Pp. 387–400 in *A Companion to Ancient Education*, edited by Martin W. Bloomer. Chichester, U.K.: Wiley-Blackwell, 2015.

Siegmund-Schultze, Reinhard. "Applied Mathematics versus Fluid Dynamics: The Catalytic Role of Richard von Mises (1883–1953)." *Historical Studies in the Natural Sciences* 48.4(2018): 475–525.

Simon, Martin A. "An Emerging Methodology for Studying Mathematics Concept Learning and Instructional Design." *The Journal of Mathematical Behavior* 52(2018): 113–21.

Sinclair, Nathalie, and Margaret Patterson. "The Dynamic Geometrisation of Computer Programming." *Mathematical Thinking and Learning* 20.1(2018): 54–74.

Singley, Alison T. Miller, and Silvia A. Bunge. "Eye Gaze Patterns Reveal How We Reason about Fractions." *Thinking and Reasoning* 24.4(2018): 445–68.

Skovsmose, Ole. "Critical Constructivism: Interpreting Mathematics Education for Social Justice." *For the Learning of Mathematics* 38.1(2018): 38–42.

Small, Marian, and Amy Lin. "Instructional Feedback in Mathematics." Pp. 169–90 in *The Cambridge Handbook of Instructional Feedback*, edited by Anastasiya A. Lipnevich and Jeffrey K. Smith. Cambridge, U.K.: Cambridge University Press, 2019.

Smith, Letisha. "Cooking Up Statistics: The Science and the Art." *Significance* 15.5(2018): 28–32.

Smith, Linda B., et al. "The Developing Infant Creates a Curriculum for Statistical Learning." *Trends in Cognitive Sciences* 22.4(2018): 325–36.

Sprenger, Jan. "Foundations of a Probabilistic Theory of Causal Strength." *Philosophical Review* 127.3(2018): 371–98.

Stark, Philip B., and Andrea Saltelli. "Cargo-Cult Statistics and Scientific Crisis." *Significance* 15.4(2018): 40–43.

Steingart, Anna. "Mathematizing." Pp. 111–18 in *Experience: Culture Cognition in the Common Sense*, edited by Caroline A. Jones, David Mather, and Rebecca Uchill. Cambridge, MA: MIT Center for Art, Science & Technology, 2016.

Strevens, Micheal. "The Mathematical Route to Causal Understanding." Pp. 96–116 in *Explanation beyond Causation: Philosophical Perspectives on Non-Causal Explanations*, edited by Alexander Reutlinger and Juha Saati. Oxford, U.K.: Oxford University Press, 2018.

Stupel, Moshe, and Victor Oxman. "Integrating Various Fields of Mathematics in the Process of Developing Multiple Solutions to the Same Problems in Geometry." *Australian Senior Mathematics Journal* 32.1(2018): 26–41.

Sun, Kathy Liu, "The Role of Mathematics Teaching in Fostering Student Growth Mindset." *Journal for Research in Mathematics Education* 49.3(2018): 330–55.

Swanson, David. "Vernacular: A Montage of Mathematical Graffiti, Quotations and Commentary." *Mathematics Enthusiast* 15.1/2(2018): 133–49.

Sword, Sarah, et al. "Leaning on Mathematical Habits of Mind." *Mathematics Teacher* 111.4 (2018): 256–63.

Tabachnikov, Serge. "A Four-Vertex Theorem for Frieze Patterns?" *The Mathematical Intelligencer* 40.4(2018): 14–18.

Taganap, Eduard C., and Ma. Louise Antonette N. de Las Peñas. "Hyperbolic Isocoronal Tilings." *Journal of Mathematics and the Arts* 12.2/3(2018): 96–110.

Tanton, James. "Just Teach My Kids the *Adjective* Math." *Medium Online* Jan. 24, 2018.

Târziu, Gabriel. "Importance and Explanatory Relevance: The Case of Mathematical Explanations." *Journal for General Philosophy of Science* 49(2018): 393–412.

Taylor, Janet A., and Michael J. Brickhill. "Enabling Mathematics: Curriculum Design to Support Transfer." *Australian Senior Mathematics Journal* 32.1(2018): 42–53.

Teran, Joseph. "Movie Animation: A Continuum Approach for Frictional Contact." *Notices of the American Mathematical Society* 65.8(2018): 909–12.

Thieme, Nick. "R Generation." *Significance* 15.4(2018): 14–19.

Thieme, Nick. "Statistics in Court." *Significance* 15.5(2018): 14–17.

Thomas, R. S. D. "An Appreciation of the First Book of Spherics." *Mathematics Magazine* 91.1(2018): 3–15.

Thompson, Katherine. "A Survey of the Math Blogosphere." *Journal of Humanistic Mathematics* 8.1(2018).

Tokhmechian, Ali, and Minou Gharehbaglou. "Music, Architecture and Mathematics in Traditional Iranian Architecture." *Nexus Network Journal* 20(2018): 353–71.

Treeby, David. "Applying Archimedes's Method to Alternating Sums of Powers." *The Mathematical Intelligencer* 40.4(2018): 65–70.

Trlifajová, Kateřina. "Bolzano's Infinite Quantities." *Foundations of Science* 23(2018): 681–704.

Tsiamis, Mia, Alfonso Oliva, and Michele Calvano. "Algorithmic Design and Analysis of Architectural Origami." *Nexus Network Journal* 20(2018): 59–73.

Tunstall, Samuel Luke. "College Algebra: Past, Present, and Future." *PRIMUS* 28.7(2018): 627–40.

Tylman, Wojciech. "Computer Science and Philosophy." *Foundations of Science* 23(2018): 159–72 [followed by a Lorenz Demey rejoinder].

Ueda, Yoshiyuki, et al. "Cultural Differences in Visual Search for Geometric Figures." *Cognitive Science* 42(2018): 286–310.

van Bommel, Martin F., and Katie T. MacEachern. "Armies of Chess Queens." *The Mathematical Intelligencer* 40.2(2018): 10–15.

van Weerden, Anne, and Steven Wepster. "A Most Gossiped about Genius: Sir William Rowan Hamilton." *BSHM Bulletin—Journal of the British Society for the History of Mathematics* 33.1(2018): 2–20.

Vasudevan, Anubav. "Chance, Determinism and the Classical Theory of Probability." *Studies in History and Philosophy of Science, A* 67(2018): 32–43.

Viana, Vera, "From Solid to Plane Tessellations, and Back." *Nexus Network Journal* 20(2018): 741–68.

Vidal, Climent J., and J. Soliveres Tur. "The Modernity of Dedekind's Anticipations Contained in *What Are Numbers and What Are They Good For?*" *Archive for History of Exact Sciences* 72(2018): 99–141.

von Plato, Jan. "In Search of the Sources of Incompleteness." Pp. 4043–61 in *Proceedings of the International Congress of Mathematics, Rio de Janeiro 2018*, edited by Boyan Sirakov, Paulo Ney de Souza, and Marcelo Viana. Singapore: World Scientific, 2018.

von Plato, Jan. "Kurt Gödel's First Steps in Logic: Formal Proofs in Arithmetic and Set Theory through a System of Natural Deduction." *Bulletin of Symbolic Logic* 24.3(2018): 319–35.

von Renesse, Christine, and Jennifer DiGrazia. "Mathematics, Writing, and Rhetoric: Deep Thinking in First-Year Learning Communities." *Journal of Humanistic Mathematics* 8.1 (2018).

Vosniadou, Stella, et al. "The Recruitment of Shifting and Inhibition in On-Line Science and Mathematics Tasks." *Cognitive Science* 42(2018): 1860–86.

Wan, Zhaoyuan. "Newton in China: Translating the *Principia* into Chinese (c. 1855–2015)." *Annals of Science* 75.1(2018): 1–20.

Wasserman, Nicholas H. "Knowledge of Nonlocal Mathematics for Teaching." *The Journal of Mathematical Behavior* 49(2018): 116–28.

Weinberg, Aaron, and Matthew Thomas. "Student Learning and Sense-Making from Video Lectures." *International Journal of Mathematical Education in Science and Technology* 49.6(2018): 922–43.

Wertheim, Margaret. "Radical Dimensions." *Aeon Online* Jan. 10, 2018.

Wiering, Marco. "Reinforcement Learning: From Methods to Applications." *Nieuw Archief voor Wiskunde* 5/19.3(2018): 157–66.

Wilkinson, Louise C. "Teaching the Language of Mathematics: What the Research Tells Us Teachers Need to Know and Do." *The Journal of Mathematical Behavior* 51(2018): 167–74.

Wolfram, Stephen. "Are All Fish the Same Shape If You Stretch Them? The Victorian Tale of *On Growth and Form*." *The Mathematical Intelligencer* 40.4(2018): 39–62.

Woods, Jack. "Mathematics, Morality, and Self-Effacement." *Noûs* 52.1(2018): 47–68.

Woolcott, Geoff. "The Tail of a Whale." *Australian Mathematics Teacher* 74.3(2018): 4–13.

Wu, Jiangmei. "Folding Helical Triangle Tessellations into Light Art." *Journal of Mathematics and the Arts* 12.1(2018): 19–33.

Yaşar, Osman. "A New Perspective on Computational Thinking." *Communications of the ACM* 61.7(2018): 33–39.

Yli-Vakkuri, Juhani, and John Hawthorne. "The Necessity of Mathematics." *Noûs* 52.1(2018): 1–28.

Yopp, David A. "When an Argument Is the Content: Rational Number Comprehension through Conversions across Registers." *The Journal of Mathematical Behavior* 50(2018): 42–56.

Young, Adena E., and Frank C. Worrell. "Comparing Metacognition Assessments of Mathematics in Academically Talented Students." *Gifted Child Quarterly* 62.3(2018): 259–75.

Zahle, Julie. "Values and Data Collection in Social Research." *Philosophy of Science* 85.1(2018): 144–63.

Zamboj, Michal. "Double Orthogonal Projection of Four-Dimensional Objects onto Two Perpendicular Three-Dimensional Spaces." *Nexus Network Journal* 20(2018): 267–81.

Zamboj, Michal. "Sections and Shadows of Four-Dimensional Objects." *Nexus Network Journal* 20(2018): 475–87.

Zhmud, Leonid, and Alexei Kouprianov. "Ancient Greek *Mathēmata* from a Sociological Perspective." *Isis* 109.3(2018): 445–72.

Ziegler, Günter M. "For Example: On Occasion of the Fiftieth Anniversary of Grünbaum's 'Convex Polytopes.'" *Notices of the American Mathematical Society* 65.5(2018): 531–36.

Zimmermann, Eckart. "Small Numbers Are Sensed Directly, High Numbers Constructed from Size and Density." *Cognition* 173(2018): 1–7.

Notable Book Reviews and Review Essays

The number of published book reviews relevant to this volume is enormous every year. The following selection is not comprehensive and not even representative but rather highly selective; it contains substantive reviews I happened to encounter while surveying the literature. My goal in providing this list is to offer yet more reading and research suggestions to the interested reader. To save space, I give the main title of the book only, no additional bibliographic information, and only the first reviewer's name (when more than one). Here *BSHM* stands for *BSHM Bulletin—Journal of the British Society for the History of Mathematics*, *NAMS* stands for *Notices of the American Mathematical Society*, and *MI* for *The Mathematical Intelligencer*.

E. Abrams reviews *American Mathematics 1890–1913* by Steve Batterson. *BSHM* 33.3(2018): 200–201.

J. Agar reviews *Reckoning with Matter* by Matthew L. Jones. *Isis* 109.3(2018): 634–35.

H. Andersen reviews *Because without Cause* by Marc Lange. *Mind* 127.506(2018): 593–602.

T. Archibald reviews *Historiography of Mathematics in the Nineteenth and Twentieth Centuries*, edited by Volker Remmert, Martina Schneider, and Henrik Kragh Sørensen. *Isis* 109.2(2018): 370–72.

A. Ash reviews *Prime Numbers and the Riemann Hypothesis* by Barry Mazur and William Stein. *American Mathematical Monthly* 125.5(2018): 476–80.

B. Assadian reviews *Abstractionism* by Philip Ebert and Marcus Rossberg. *Analysis* 78(2018): 188–91.

T. Banchoff reviews *The Seduction of Curves* by Allan McRobie. *Journal of Mathematics and the Arts* 12.4(2018): 252–56.

D. Beckers reviews *Mathematical Cultures*, edited by Brendan Larvor. *Isis* 109.2(2018): 365–66.

D. Bellhouse reviews *The Life and Works of John Napier* by Brian Rice, Enrique González-Velasco, and Alexander Corrigan. *Isis* 109.2(2018): 396–97.

C. Besson reviews *Lewis Carroll's Paradox of Inference*, edited by Amirouche Moktefi and Francine Abeles. *History and Philosophy of Logic* 39.1(2018): 96–98.

J. Best reviews *The Tyranny of Metrics* by Jerry Muller. *Numeracy* 11.2(2018).

M. Blanco reviews *Magic Squares in the Tenth Century* by Jacques Sesiano. *Isis* 109.2(2018): 381–82.

R. Cook reviews *Essays on Paradoxes* by Terence Horgan. *Analysis* 78(2018): 567–69.

R. Cowen reviews *A Beginner's Guide to Mathematical Logic* and *A Beginner's Further Guide to Mathematical Logic* by Raymond Smullyan. *American Mathematical Monthly* 125.2(2018): 188–92.

T. Crilly reviews *Ten Great Ideas about Chance* by Persi Diaconis and Brian Skyrms. *BSHM* 33.3(2018): 197–99.

M. Danino reviews *Les Mathématiques de l'Autel Védique* by Jean-Michel Delire. *Isis* 109.1(2018): 163–64.

J. W. Dawson Jr. reviews *The Great Formal Machinery Works* by Jan von Plato. *BSHM* 33.3(2018): 196–97.

G. Dietz reviews *The Mathematics Lover's Companion* by Edward Scheinerman. *American Mathematical Monthly* 125.1(2018): 90–95.

S. Ducheyne reviews *Thomas Reid on Mathematics and Natural Philosophy*, edited by P. Wood. *Annals of Science* 75.4(2018): 369–71.

C. M. Eddy reviews *Sociopolitical Dimensions of Mathematics Education*, edited by Murad Jurdak and Renuka Vithal. *Educational Studies in Mathematics* 99(2018): 359–62.

T. Edgar reviews Roger Nelson's books "so far." *College Mathematics Journal* 49.4(2018): 302–12.

D. Faraci reviews *Explanation in Ethics and Mathematics* by Uri Leibowitz and Neil Sinclair. *Analysis* 78(2018): 377–81.

F. Feather reviews *The Calculus Story* by David Acheson. *BSHM* 33.3(2018): 192–94.

J. Ferreirós reviews *Making and Breaking Mathematical Sense* by Roi Wagner. *Philosophia Mathematica* 26.1(2018): 131–48.

M. Forgione reviews *Because without Cause* by Marc Lange. *Journal for General Philosophy of Science* 49(2018): 487–90.

M. Fried reviews *Closing the Gap* by Vicky Neale. *Mathematical Thinking and Learning* 20.3(2018): 248–50.

P. Fritz reviews *The Boundary Stones of Thought* by Ian Rumfitt. *Mind* 127.505(2018): 265–76.

D. Glass reviews *The Joy of SET* by Liz McMahon et al. *American Mathematical Monthly* 125.3 (2018): 284–88.

D. Glass reviews *Unsolved!* by Craig Bauer. *Math Horizons* 25.3(2018): 28.

J. Golden reviews *Geometry Snacks* by Ed Southall and Vincent Pantaloni. *Math Horizons* 26.2(2018): 28.

M. W. Gray reviews *A Lady Mathematician in This Strange Universe* by Yvonne Choquet-Bruhat. *MI* 40.4(2018): 86–88.

N. Guicciardini reviews *De quadratura arithmetica circuli ellipseos et hyperbolae cujus corollarium est trigonometria sine tabulis* by Gottfried Wilhelm Leibniz, edited by Eberhard Knobloch. *MI* 40.4(2018): 89–91.

A. Haensch reviews *Foolproof and Other Mathematical Meditations* by Brian Hayes. *Math Horizons* 25.4(2018): 29.

K. H. Hamman reviews *Curbing Catastrophe* by Timothy Dixon. *Numeracy* 11.2(2018).

K. H. Hamman reviews *A Numerate Life* by John Allen Paulos. *Numeracy* 11.1(2018).

J. Hansen reviews *Visible Numbers*, edited by Miles Kimball and Charles Kostelnick. *Isis* 109.1(2018): 150–51.

A. Heefer reviews *Problems for Metagrobologists* by David Singmaster. *MI* 40.4(2018): 92–93.

C. D. Hollings reviews *The Case of Academician Nikolai Nikolaevich Luzin* by Sergei S. Demidov and Boris V. Lëvshin. *BSHM* 33.1(2018): 59–61.

N. M. Joseph reviews *Cases for Mathematics Teacher Educators* edited by Dorothy Y. White, Sandra Crespo, and Marta Civil. *Journal for Research in Mathematics Education* 49.2(2018): 232–36.

B. P. Katz reviews *A TEXas-Style Introduction to Proof* by Ron Taylor and Patrick X. Rault; and *Discovering Discrete Dynamical Systems* by Aimee S. A. Johnson, Kathleen M. Madden, and Ayşe A. Şahin. *College Mathematics Journal* 49.5(2018): 378–83.

J. Kilpatrick reviews *Compendium for Research in Mathematics Education*, edited by Jinfa Cai. *Journal for Research in Mathematics Education* 49.4(2018): 501–12.

R. Kossak reviews *Abstraction and Infinity* by Paolo Mancosu. *MI* 40.2(2018): 89–90.

B. Larvor reviews *Making and Breaking Mathematical Sense* by Roi Wagner. *NAMS* 65.6(2018): 692–95.

G. Lavers reviews *Philosophy of Mathematics* by Øystein Linnebo. *Philosophia Mathematica* 26.3(2018): 413–21.

J. Makansi reviews *Curbing Catastrophe* by Timothy Dixon. *Numeracy* 11.2(2018).

D. May reviews *Ode to Numbers* by Sarah Glaz. *Journal of Mathematics and the Arts* 12.4(2018): 248–52.

L. McDonald reviews *Wonders beyond Numbers* by Johnny Ball. *BSHM* 33.3(2018): 201–3.

A. Meadows reviews *Writing Proofs in Analysis* by Jonathan M. Kane. *American Mathematical Monthly* 125.7(2018): 669–72.

T. Mormann reviews *Space, Number, and Geometry from Helmholtz to Cassirer* by Francesca Biagioli. *Philosophia Mathematica* 26.2(2018): 287–92.

C. Mummert reviews *Reverse Mathematics* by John Stillwell. *NAMS* 65.9(2018): 1098–102.

E. Neuenschwander reviews *Sourcebook in the Mathematics of Medieval Europe and North Africa*, edited by Victor J. Katz et al. *Isis* 109.1(2018): 171–73.

L. N. Oláh reviews *Psychometric Methods in Mathematics Education*, edited by Andrew Izsák, Janine Remillard, and Jonathan Templin. *Journal for Research in Mathematics Education* 49.3(2018): 356–60.

D. O'Shea reviews *Prime Numbers and the Riemann Hypothesis* by Barry Mazur and William Stein. *NAMS* 65.7(2018): 811–15.

M. Ossendrijver reviews *The Circulation of Astronomical Knowledge in the Ancient World*, edited by John M. Steele. *Journal for the History of Astronomy* 49(2018): 125–29.

A. Pais reviews *Humanizing Mathematics and Its Philosophy*, edited by Bharath Sriraman. *Educational Studies in Mathematics* 99(2018): 235–40.

T. J. Pfaff reviews *Painting by Numbers* by Jason Makansi. *Numeracy* 11.1(2018).

M. Povich reviews *Because without Cause* by Marc Lange. *Philosophical Review* 127(2018): 422–26.

P. Raatikainen reviews *Gödel's Disjunction*, edited by Leon Horsten and Philip Welch. *History and Philosophy of Logic* 39.4(2018): 401–3.

J. T. Remillard reviews *Curricular Resources and Classroom Use* by Gabriel Styliandes. *Journal for Research in Mathematics Education* 49.2(2018): 228–31.

D. Richeson reviews *The Art and Craft of Geometric Origami* by Mark Bolitho. *Math Horizons* 25.4(2018): 28.

J. Rodal reviews *Literature after Euclid* by Matthew Wickman. *BSHM* 33.1(2018): 55–57.

W. T. Ross reviews *Creating Symmetry* by Frank A. Farris. *Journal of Mathematics and the Arts* 12.1(2018): 55–58.

J. Sebestik reviews *Bolzano's Logical System* by Ettore Casari. *History and Philosophy of Logic* 39.2(2018): 164–86.

N. Sinclair reviews *Research for Educational Change*, edited by Jill Adler and Anna Sfard. *Mathematical Thinking and Learning* 20.2(2018): 162–66.

C. Smoriński reviews *The Great Formal Machinery Works* by Jan von Plato. *MI* 40.4(2018): 94–96.

D. Spencer reviews *4 3 2 1: A Novel* by Paul Auster. *MI* 40.1(2018): 80–83.

B. Sriraman reviews *The Disorder of Mathematics Education*, edited by Hauke Staehler-Pohl, Nina Bohlmann, and Alexandre Pais. *Educational Studies in Mathematics* 97(2018): 209–13.

J. Stén reviews *Leonhard Euler* by Ronald Calinger. *MI* 40.2(2018): 93–97.

P. Sztajn reviews *Research for Educational Change*, edited by Jill Adler and Anna Sfard. *Journal for Research in Mathematics Education* 49.5(2018): 614–18.

L. Taalman reviews *Visualizing Mathematics with 3D Printing* by Henry Segerman and *Pasta by Design* by George Legendre. *American Mathematical Monthly* 125.4(2018): 379–84.

S. Tabachnikov reviews *A Singular Mathematical Promenade* by Étienne Ghys. *MI* 40.2(2018): 85–88.

A. Trabesinger reviews *Mathematics and Art* by Lynn Gamwell. *Nature Physics* 13(2017): 6.

S. L. Tunstall reviews *The Improbability Principle* by David Hand and *The Logic of Miracles* by Lásló Mérő. *Numeracy* 11.2(2018).

S. L. Tunstall reviews *Weapons of Math Destruction* by Cathy O'Neil. *Numeracy* 11.1(2018).

J. von Plato reviews [Gödel's] *Logic Lectures*, edited by Miloš Adžić and Kosta Došen. *History and Philosophy of Logic* 39.4(2018): 396–401.

J. von Plato reviews *Martin Davis on Computability, Computational Logic, and Mathematical Foundations*, edited by E. Omodeo and A. Policriti. *MI* 40.2(2018): 82–84.

D. Voytenko reviews *Painting by Numbers* by Jason Makansi. *Numeracy* 11.1(2018).

R. Wagner reviews *Mathematical Knowledge and the Interplay of Practices* by José Ferreirós. *BSHM* 33.2(2018): 1411–43.

J. Watkins reviews *Mathematics without Apologies* by Michael Harris. *MI* 40.4(2018): 97–98.

R. Wilson reviews *Leonhard Euler* by Ronald Calinger. *BSHM* 33.1(2018): 54–55.

B. Wolf reviews *Significant Figures* by Ian Stewart. *Math Horizons* 25.3(2018): 29.

Teaching Tips and Notes

To save space on this list, in some cases we give only the name of the first author; the abbreviation *CMJ* stands for *College Mathematics Journal* and *IJME*, for *International Journal of Mathematical Education*.

Akhtyamov, A. "On Reconstruction of a Matrix by Its Minors." *IJME* 49.2(2018): 268–81.

Buchbinder, O. "Guided Discovery of the Nine-Point Circle Theorem and Its Proof." *IJME* 49.1(2018): 138–53.

Cunningham, D. "Why Does Trigonometric Substitution Work?" *IJME* 49.4(2018): 588–93.

de Camargo, A. "The Geometric Mean Value Theorem." *IJME* 49.4(2018): 613–15.

Diamond, H. "The Rational Approximation of Small Angles." *CMJ* 49.1(2018): 57–59.

Ferrarello, D. "Magic of Centroids." *IJME* 49.4(2018): 628–41.

García-Caballero, M. E. "The Double-Sidedness of Matrix Inverses: Yet Another Proof." *CMJ* 49.2(2018): 136–37.

Gergu, M. "Iterated Exponential Inequalities." *IJME* 49.5(2018): 802–7.

Hoseana, J. "On Zero-Over-Zero Form Limits of a Special Type." *CMJ* 49.3(2018): 219–21.

Johansson, T. "An Elementary Algorithm to Evaluate Trigonometric Functions to High Precision." *IJME* 49.1(2018): 131–37.

Kossek, W. "Is a Taylor Series Also a Generalized Fourier Series?" *CMJ* 49.1(2018): 54–56.

Lee, W.-S. "Tensor Calculus: Unlearning Vector Calculus." *IJME* 49.2(2018): 293–304.

Libeskind, S. "The Concept of Invariance in School Mathematics." *IJME* 49.1(2018): 107–20.

Lord, N. "An Unusual Inequality Challenge." *Mathematical Gazette* 102(2018): 155–59.

Lord, N. "Extending Runs of Composite Numbers." *Mathematical Gazette* 102(2018): 351–52.

Lord, N. "Three Cameos Involving *e*." *Mathematical Gazette* 102(2018): 159–64.

McDevitt, T. "A Simple Probability Paradox." *CMJ* 49.4(2018): 292–94.

Moreno-Armella, L. "Dynamic Hyperbolic Geometry." *IJME* 49.4(2018): 594–612.

Pruitt, K. "Modular Class Primes in the Sundaram Sieve." *IJME* 49.6(2018): 944–47.

Pujol, J. "The Linear Combination of Vectors Implies the Existence of the Cross and Dot Products." *IJME* 49.5(2018): 778–92.

Richmond, T. "Calculus with Curtains." *CMJ* 49.5(2018): 369–70.

Sawhney, M. "An Unusual Proof of the Triangle Inequality." *CMJ* 49.3(2018): 218.

Sinitsky, I. "Pizza Again?" *IJME* 49.2(2018): 281–93.

Tisdell, C. "Pedagogical Alternatives for Triple Integrals." *IJME* 49.5(2018): 792–801.

Treviño, E. "An Inclusion-Exclusion Proof of Wilson's Theorem." *CMJ* 49.5(2018): 367–68.

Wares, A. "An Interesting Property of Hexagons." *IJME* 49.3(2018): 437–41.

Wares, A. "Dynamic Geometry as a Context for Exploring Conjectures." *IJME* 49.1(2018): 153–59.

Notable Journal Issues

This list is ordered alphabetically by the unabbreviated title of the publication.

"Data Science." *The American Statistician* 72.1(2018).

"Human Rights." *Chance* 31.1(2018).

"Defense and National Security." *Chance* 31.2(2018).

"Sports." *Chance* 31.3(2018).

"Literature and Science." *Configurations* 26.3(2018).

"Interplay between Mathematical Journals on Various Scales, 1850–1950." *Historia Mathematica* 45.4(2018).

"History and Philosophy of Logical Notation." *History and Philosophy of Logic* 39.1(2018).

"The History of Design in Computing." *IEEE Annals of the History of Computing* 40.1(2018).

"The Local and the Global in the History of Computing." *IEEE Annals of the History of Computing* 40.2(2018).

"Science and Mathematics Literacy: PISA for Better School Education." *International Journal of Science and Mathematics Education* 16.1 suppl(2018).

"Replication in Educational Research." *Journal for Research in Mathematics Education* 49.1(2018).

"Mathematics and Motherhood." *Journal of Humanistic Mathematics* 8.2(2018).

"Pluralism in Mathematics." *Journal of Indian Council of Philosophical Research* 34.2(2017).

"An International Perspective on Knowledge in Teaching Mathematics." *The Journal of Mathematical Behavior* 49.3(2018).

"Learning through Activity: Studying and Promoting Reflective Abstraction of Mathematical Concepts." *The Journal of Mathematical Behavior* 49.4(2018).

"Computational Thinking and Mathematics Learning." *Mathematical Thinking and Learning* 20.1(2018).

"Strategic Research Directions: Large-Scale Funded Projects in Australasia." *Mathematics Education Research Journal* 30.1(2018).

"The Disorder of Mathematics Education." *Mathematics Enthusiast* 15.1–2(2018).

"Nurturing Persistent Problem Solvers." *Mathematics Teacher* 111.3(2017).

"Stereotomy 2.0." *Nexus Network Journal* 20.3(2018).

"[Richard] Feynman's Diagrams." *Perspectives on Science* 26.4(2018).

"Descartes and the First Cartesians." *Perspectives on Science* 26.5(2018).

"Aesthetics in Mathematics." *Philosophia Mathematica* 26.2(2018).

"Project-Based Curriculum." *PRIMUS* 28.4(2018).

"Improving the Teaching and Learning of Calculus." *PRIMUS* 28.6(2018).

"Hermann Weyl and the Philosophy of the 'New Physics.'" *Studies in History and Philosophy of Science, B* 61(2018).

"Mechanistic and Topological Explanations." *Synthese* 195.1(2018).

"Logic, Rationality and Interaction." *Synthese* 195.10(2018).

"Cartesian Epistemology." *Synthese* 195.11(2018).

"Preparing to Teach Mathematics Pathways beyond 16." *Teaching Mathematics and Its Applications* 37.2(2018).

"Reasoning and Mathematics." *Thinking and Reasoning* 24.2(2018).

"Empirical Research on the Teaching and Learning of Mathematical Modeling." *Zentralblatt für Didaktik der Mathematik* 50.1–2(2018).

"Studying Instructional Quality in Mathematics through Different Lenses." *Zentralblatt für Didaktik der Mathematik* 50.3(2018).

"Assessment in Mathematics Education: Issues Regarding Methodology, Policy and Equity." *Zentralblatt für Didaktik der Mathematik* 50.4(2018).

"Recent Advances in Mathematics Textbook Research and Development." *Zentralblatt für Didaktik der Mathematik* 50.5(2018).

"Research Frameworks for the Study of Language in Mathematics Education." *Zentralblatt für Didaktik der Mathematik* 50.6(2018).

"Innovations in Statistical Modelling to Connect Data, Chance and Context." *Zentralblatt für Didaktik der Mathematik* 50.7(2018).

Acknowledgments

The authors of the pieces in this volume and the original publishers deserve my thanks for their permissions, cooperation, and solicitude during the preparation of the book.

Many editors at various publishing houses ignored or rejected my proposals to start an annual series of anthologies of writings on mathematics—until Steven Strogatz put me in contact with Vickie Kearn, ten years ago. This series would not have existed without Vickie's patience, understanding, enthusiasm, guidance, and determination to make it happen. I dedicate to her this volume—the last to which Vickie contributed her in-house editorial work, before retiring. One of our common friends once told me that over the past three decades Vickie has been "a force" in mathematics publishing: A moving and positive force, indeed. Thank you, Vickie!

On the publishing side, I also thank Susanna Shoemaker for taking over Vickie's duties in a most able manner; Nathan Carr for yet another feat of excellent work in making sure that the book was ready for production; and Paula Bérard for her careful work as a copyeditor. Also thanks to Patrick Hancy of the Cornell University Library for a useful last-moment piece of information.

A strange thing happened to me over the past two years: I made a living by doing what I can do best, teaching mathematics. It took me 25 years of epic failures and misadventures to get here; I was told many times that I could not teach mathematics in schools because I lack "certification"—something that (as it turns out) has nothing to do with the ability to teach mathematics. During those years, I survived by learning and practicing other skills, some useful to me to this day. Many thanks to Leonid Kovalev and Jeffrey Meyer for assigning me an adequate teaching load and a convenient schedule of classes in the Department of Mathematics at Syracuse University; to Uday Banerjee for renewing my employment contract; and to Kelly Jarvi, Julia O'Connor, Leah Quinones, and Jordan Dias Correia for assisting in various administrative matters. They all made it possible for me to minimize commuting time from Ithaca to Syracuse and enabled me to direct my attention beyond the minimal goal of mere survival.

Raising two toddlers and a teenager in an actively trilingual household is an exciting and challenging experience. It makes for an environment not exactly conducive to quiet meditation. My family supported me as much as I can hope for in such circumstances—and I thank them all with affection.

Credits

"Geometry v. Gerrymandering" by Moon Duchin. Originally published in *Scientific American* 319.5(2018): 49–53. Reprinted by permission.

"Slicing Sandwiches, States, and Solar Systems: Can Mathematical Tools Help Determine What Divisions are Provably Fair?" by Theodore P. Hill. Originally published in *The American Scientist* 106.1(Jan.–Feb. 2018): 42–49. Reprinted by permission of Sigma Xi, The Scientific Research Society.

"Does Mathematics Teach How to Think?" by Paul J. Campbell. Partly based on articles by Paul J. Campbell published in *The UMAP Journal of Undergraduate Mathematics and Its Applications* (2006, 2016, 2017, and 2018). Reprinted by permission of Consortium of Mathematics and Its Applications.

"Abstracting the Rubik's Cube" by Roice Nelson. Originally published in *Math Horizons* 25.4(2018): 18–22. Copyright © 2018 Mathematical Association of America, https://www.maa.org/, reprinted by permission of Taylor & Francis Ltd, http://www.tandfonline.com, on behalf of the Mathematical Association of America.

"Topology-Disturbing Objects: A New Class of 3D Optical Illusion" by Kokichi Sugihara. Originally published in *Journal of Mathematics and the Arts* 12.1(2018): 2–18. Copyright © 2018 Taylor & Francis Ltd, http://www.tandfonline.com.

"Mathematicians Explore Mirror Link between Two Geometric Worlds" by Kevin Hartnett. Originally published on April 4, 2018, in *Quanta Magazine Online*. Original story reprinted with permission from Quanta Magazine (www.quantamagazine.org) an editorially independent publication of the Simons Foundation whose mission is to enhance public understanding of science by covering research developments and trends in mathematics and the physical and life sciences.

"Professor Engel's Marvelously Improbable Machines" by James Propp. Originally published in *Math Horizons* 26.2(2018): 5–9. Copyright © 2018 Mathematical Association of America, https://www.maa.org/, reprinted by permission of Taylor & Francis Ltd, http://www.tandfonline.com, on behalf of the Mathematical Association of America.

"The On-Line Encyclopedia of Integer Sequences" by Neil J. A. Sloane. Originally published in *Notices of the American Mathematical Society* 65.9(2018): 1062–74. Copyright © 2018 American Mathematical Society. Reprinted by permission of the Royal Mathematical Society.

"Mathematics for Big Data" by Alessandro Di Bucchianico, Laura Iapichino, Nelly Litvak, Frank van der Meulen, and Ron Wehrens. Originally published in *Nieuw Archief voor Wiskunde* 19.4(2018): 282–86. Reprinted by permission of the Royal Mathematical Society.

"The Un(solv)able Problem" by Toby S. Cubitt, David Pérez-García, and Michael Wolf. Originally published in *Scientific American* 319.4(2018): 29–37. Reprinted by permission of Springer Nature.

"The Mechanization of Mathematics" by Jeremy Avigad. Originally published in *Notices of the American Mathematical Society* 65.6(2018): 681–90. Copyright © 2018 American Mathematical Society. Reprinted by permission.

"Mathematics as an Empirical Phenomenon, Subject to Modeling" by Reuben Hersh. Originally published in *Journal of Indian Council of Philosophical Research* 34.2(2017): 331–42. Reprinted by permission of Springer Nature.

"Does 2 + 3 = 5? In Defence of a Near Absurdity" by Mary Leng. Originally published in *The Mathematical Intelligencer* 40.1(March 2018): 14–17. Reprinted by permission of Springer Nature.

"Gregory's Sixth Operation" by Tiziana Bascelli, Piotr Błaszczyk, Vladimir Kanovei, Karin U. Katz, Mikhail G. Katz, Semen S. Kutateladze, Tahl Nowik, David M. Schaps, and David Sherry. Originally published in *Foundations of Science* 23(2018): 133–44. Reprinted by permission of Springer Nature.

"Kolmogorov Complexity and Our Search for Meaning: What Math Can Teach Us about Finding Order in our Chaotic Lives" by Noson S. Yanofsky. Originally published in *Nautilus Online* Aug. 2, 2018. Reprinted by permission of Nautilus Think.

"Ethics in Statistical Practice and Communication: Five Recommendations" by Andrew Gelman. Originally published in *Significance* 15.5(2018): 40–43. Reprinted by permission of John Wiley & Sons.

"The Fields Medal Should Return to Its Roots" by Michael J. Barany. Originally published in *Nature* 553 (Jan. 12, 2018): 271–73. Reprinted by permission of Springer Nature.

"The Erdős Paradox" by Melvyn B. Nathanson. Originally plublished in *Combinatorial and Additive Number Theory II*, edited by Melvyn B. Nathanson. Cham, Switzerland: Springer Nature, 2017, pp. 249–54. Reprinted by permission of Springer Nature.